Renewable Motor Fuels

Renewable Heating Fuel

Renewable Motor Fuels:

The Past, the Present and the Uncertain Future

Arthur M. Brownstein
ExxonMobil Corp., Retired

AMSTERDAM • BOSTON • HEIDELBERG • LONDON
NEW YORK • OXFORD • PARIS • SAN DIEGO
SAN FRANCISCO • SINGAPORE • SYDNEY • TOKYO
Butterworth-Heinemann is an imprint of Elsevier

Butterworth-Heinemann is an imprint of Elsevier
The Boulevard, Langford Lane, Kidlington, Oxford, OX5 1GB, UK
225 Wyman Street, Waltham, MA 02451, USA

Library of Congress Cataloging-in-Publication Data
A catalog record for this book is available from the Library of Congress

British Library Cataloguing-in-Publication Data
A catalogue record for this book is available from the British Library

ISBN: 978-0-12-800970-3

For information on all Butterworth-Heinemann publications
visit our website at **http://store.elsevier.com**

This book has been manufactured using Print On Demand technology. Each copy is produced to
order and is limited to black ink. The online version of this book will show color figures where
appropriate.

Working together
to grow libraries in
developing countries

www.elsevier.com • www.bookaid.org

CONTENTS

Where do renewable fuels go from here? To answer this, it's important to know not only where they have been but where they are presently. George Santanyana, the philosopher, is noted for having said, "those who cannot remember the past are condemned to repeat it". This is a cautionary note that is pertinent to legislation and to technology development for renewable fuels.

In 2005, Congress determined that fuels produced from renewable resources, i.e., biofuels, should play a major role in ensuring US energy independence. A collateral objective was a decrease in greenhouse gas emissions. This legislation was enacted at a time when US reserves of natural gas and crude oil were seen to be declining and global warming was on the rise.

The 2005 biofuels legislation had the Renewable Fuels Standard (RFS) as a centerpiece. This stipulated that a minimum volume of biofuels by type, such as ethanol, must be used annually in US motor fuels. The basis for the RFS was the Energy Policy Act of 2005 that provided tax incentives and loan guarantees for innovative energy development of various types. The Act of 2005 was quickly followed by the Energy Independence and Security Act of 2007 (EISA). This second Act extended the domestic biofuels target to 36 billion gallons by 2022, established new categories of renewable fuels and set minimum targets for each.

The legislative rush to renewable fuels to solve a perceived energy shortage resulted in a boom in cornstarch ethanol fermentation. The quick decision to focus on cornstarch ethanol fermentation was to be expected since its technology was well established and corn supplies were abundant. Congress was also well aware of the success of Brazil who had developed energy independence based on the successful fermentation of their sugar cane to fuel ethanol. In the years following 2002, US production of fuel grade ethanol soared from 2 to 14 billion gallons. Motor fuel quickly commanded 40% of US corn consumption and food prices rose accordingly. This was the unintended consequences of the 2005–2007 legislation. To resolve an untenable situation, EPA pressed fuel ethanol producers to use cellulosic waste materials, in addition to, or in place of cornstarch. The producers were initially fined in 2012 for noncompliance for the use of cellulose technology that was clearly only in the early stages of its development and unavailable commercially. The fines were rescinded after a court action, and efforts to develop the technology are continuing. In the meantime, growth in cornstarch fermentation has declined.

There are two general avenues of approach in the use of waste cellulose. One is to ferment the cellulose directly, and the other is to first convert it to a carbon monoxide/hydrogen synthesis gas which is then fermented to fuels. Concurrently, there are efforts to convert cornstarch and cellulose fermentation plants to produce isobutanol in place of ethanol. As a motor fuel, isobutanol has a number of advantages over ethanol and could conceivably replace it entirely. The synthesis gas based processes that are able to produce isobutanol can also produce diesel fuel. Both isobutanol and diesel, when based on biomass, qualify under the EISA legislation as a renewable fuel. Some major oil companies, expect biodiesel to surpass gasoline as the primary motor fuel by 2020. In the view of the US Energy Information Agency, diesel and isobutanol may account for the majority of all renewable fuels by 2022.

Natural gas has hovered in the background of all renewable fuels development since 2005. It plays a competitive as well as a complementary role in processes under development. The advent of fracking was almost simultaneous with the introduction of the Renewable Fuels Standard. The subsequent enormous growth in natural gas reserves could only serve to slow the growth of renewable fuels because it undercuts the urgency of the strategy enacted in 2005. At the same time, many developers of renewable fuels have looked to natural gas as the source of the synthesis gas that they hope to use. In this instance, the resulting biofuels are a combination of the two technologies. There is also an expectation that heavy duty vehicles and passenger cars may be powered by natural gas itself. Finally, there is the electric automobile, whether as a hybrid or as a battery or as a fuel cell powered vehicle. The fuels used in such hybrids could be renewable as could the fuel to generate the electricity for them at the power plant itself.

Billions of dollars have been expended in federal, state, and private funding to develop a large number of renewable fuels technology options. Many of these expenditures have gone for naught. There will be a few winners until history repeats itself and introduces issues not yet contemplated. The history of automobile fuels is littered with the bodies of past innovations. Diesel was introduced in the nineteenth century, then abandoned and later rediscovered. The electric automobile was the primary means of powering a vehicle in the early twentieth century, then abandoned and most recently rediscovered. Coal liquefaction was once seriously considered as a response to the Arab and OPEC oil embargoes, and after expensive development, was discarded in the early 1980s. And finally, not too long ago, after a century of use, lead that was used to enhance the performance of gasoline was found to be harmful and banished.

Where do renewable fuels go from here? To answer this, it's important to know not only where they have been but where they are presently. George Santanyana, the philosopher, is noted for having said, "those who cannot remember the past are condemned to repeat it". This is a cautionary note that is pertinent to legislation and to technology development for renewable fuels.

In 2005, Congress determined that fuels produced from renewable resources, i.e., biofuels, should play a major role in ensuring US energy independence. A collateral objective was a decrease in greenhouse gas emissions. This legislation was enacted at a time when US reserves of natural gas and crude oil were seen to be declining and global warming was on the rise.

The 2005 biofuels legislation had the Renewable Fuels Standard (RFS) as a centerpiece. This stipulated that a minimum volume of biofuels by type, such as ethanol, must be used annually in US motor fuels. The basis for the RFS was the Energy Policy Act of 2005 that provided tax incentives and loan guarantees for innovative energy development of various types. The Act of 2005 was quickly followed by the Energy Independence and Security Act of 2007 (EISA). This second Act extended the domestic biofuels target to 36 billion gallons by 2022, established new categories of renewable fuels and set minimum targets for each.

The legislative rush to renewable fuels to solve a perceived energy shortage resulted in a boom in cornstarch ethanol fermentation. The quick decision to focus on cornstarch ethanol fermentation was to be expected since its technology was well established and corn supplies were abundant. Congress was also well aware of the success of Brazil who had developed energy independence based on the successful fermentation of their sugar cane to fuel ethanol. In the years following 2002, US production of fuel grade ethanol soared from 2 to 14 billion gallons. Motor fuel quickly commanded 40% of US corn consumption and food prices rose accordingly. This was the unintended consequences of the 2005–2007 legislation. To resolve an untenable situation, EPA pressed fuel ethanol producers to use cellulosic waste materials, in addition to, or in place of cornstarch. The producers were initially fined in 2012 for noncompliance for the use of cellulose technology that was clearly only in the early stages of its development and unavailable commercially. The fines were rescinded after a court action, and efforts to develop the technology are continuing. In the meantime, growth in cornstarch fermentation has declined.

There are two general avenues of approach in the use of waste cellulose. One is to ferment the cellulose directly, and the other is to first convert it to a carbon monoxide/hydrogen synthesis gas which is then fermented to fuels. Concurrently, there are efforts to convert cornstarch and cellulose fermentation plants to produce isobutanol in place of ethanol. As a motor fuel, isobutanol has a number of advantages over ethanol and could conceivably replace it entirely. The synthesis gas based processes that are able to produce isobutanol can also produce diesel fuel. Both isobutanol and diesel, when based on biomass, qualify under the EISA legislation as a renewable fuel. Some major oil companies, expect biodiesel to surpass gasoline as the primary motor fuel by 2020. In the view of the US Energy Information Agency, diesel and isobutanol may account for the majority of all renewable fuels by 2022.

Natural gas has hovered in the background of all renewable fuels development since 2005. It plays a competitive as well as a complementary role in processes under development. The advent of fracking was almost simultaneous with the introduction of the Renewable Fuels Standard. The subsequent enormous growth in natural gas reserves could only serve to slow the growth of renewable fuels because it undercuts the urgency of the strategy enacted in 2005. At the same time, many developers of renewable fuels have looked to natural gas as the source of the synthesis gas that they hope to use. In this instance, the resulting biofuels are a combination of the two technologies. There is also an expectation that heavy duty vehicles and passenger cars may be powered by natural gas itself. Finally, there is the electric automobile, whether as a hybrid or as a battery or as a fuel cell powered vehicle. The fuels used in such hybrids could be renewable as could the fuel to generate the electricity for them at the power plant itself.

Billions of dollars have been expended in federal, state, and private funding to develop a large number of renewable fuels technology options. Many of these expenditures have gone for naught. There will be a few winners until history repeats itself and introduces issues not yet contemplated. The history of automobile fuels is littered with the bodies of past innovations. Diesel was introduced in the nineteenth century, then abandoned and later rediscovered. The electric automobile was the primary means of powering a vehicle in the early twentieth century, then abandoned and most recently rediscovered. Coal liquefaction was once seriously considered as a response to the Arab and OPEC oil embargoes, and after expensive development, was discarded in the early 1980s. And finally, not too long ago, after a century of use, lead that was used to enhance the performance of gasoline was found to be harmful and banished.

Arthur M. Brownstein
June 2014

ACKNOWLEDGMENT

To Rosalie for her forbearance, insight, and literary help.

CHAPTER 1

History and Legislation

1.1 PAST

The concept of renewable fuels in motor vehicles is not new and dates back more than 100 years to 1896 and the quadricycle designed and built by Henry Ford. This was a modified four-wheeled bicycle that had a 4 horsepower, 2 cylinder engine powered by pure ethanol. The Standard Oil Company of New Jersey (ESSO) began adding ethanol to its gasoline in the 1920s to increase the octane number and to decrease knocking. The US Army built and operated the first fuel ethanol plant in the same decade.

Concurrent with efforts in the United States, Rudolf C.K. Diesel in Europe in 1892 invented the engine that bears his name. The diesel engine was originally designed to run on coal dust. Since the petroleum industry was then in its infancy, Diesel experimented with engine modifications that permitted the use of vegetable oil as fuel. As the petroleum industry grew, a fuel was developed that enabled its use in an automobile in 1927. This was followed by the use of diesel fuel in the first production automobile, the Mercedes-Benz 206 in 1936.

Volkswagen introduced the first compact car, the Diesel-Jetta, in 1975. This was the most fuel efficient car at the time. Because diesel fuels initially had a high sulfur content, many nations, including the United States, banned its use in passenger cars because of pollution. In 2006, the Environmental Protection Agency (EPA) mandated the use of low sulfur diesel fuel in the United States. With minor engine modifications, most modern diesel engines can use waste vegetable oil directly by simply filtering out any solid impurities. In a sense, this returns us to Rudolf Diesel's original efforts of more than 100 years ago. This is a commonplace practice and is pursued by many do-it-yourself innovators.

In 2005, the US Congress determined that fuels produced from renewable resources, i.e., biofuels, would play a major role in ensuring US energy independence. A collateral objective was an effort to decrease

Table 1.1 RFS2 Schedule Under EISA of 2007 (billion gallons)			
Year	Conventional Biofuels	Advanced Biofuels	Total Renewables
2012	13.2	2.0	15.2
2013	13.8	2.75	16.55
2015	15.0	5.5	20.5
2017	15.0	9.0	24.0
2022	15.0	21.0	36.0

greenhouse gas emissions. The legislation was enacted at a time when US reserves of natural gas and crude oil were seen to be in decline and imports were dramatically increasing. As will be noted later in this chapter, the US energy situation changed radically only 5 years later.

The 2005 biofuels Congressional legislation had the Renewable Fuels Standard (RFS) as a centerpiece. This stipulated that a minimum volume of biofuels, such as ethanol, must be used annually in US motor fuels. The basis for the RFS was the Energy Policy Act of 2005 that provided tax incentives and loan guarantees for innovative energy production of various types. This was the most recent of various abortive Congressional mandates that date back to the Carter and Nixon administrations. The Nixon administration sought to ensure that gasoline prices would never exceed $1 per gallon while the Carter administration endeavored to determine that the United States would never again import as much oil as it had in 1977. History shows that neither of these well-meaning efforts was even remotely effective.

The Energy Policy Act of 2005 increased the minimum amount of biofuels that must be added to gasoline to a total of 4 billion gallons by 2006 and then to 7.5 billion gallons by 2012. This was quickly followed by the Energy Independence and Security Act (EISA) of 2007 that extended the biofuels target to 36 billion gallons by 2020. The 2007 Act also established new categories of renewable fuels and set minimum requirements for each as a revised Renewable Fuels Standard (RFS2) as shown in Table 1.1.

The EISA of 2007 was the first to require the creation of a diesel fuel from biomass provided that it can reduce greenhouse gas emissions by at least 50% compared to petroleum-derived diesel. Under the Act, conventional biofuels as a category (Table 1.1) refer to ethanol produced by fermentation of cornstarch, and the RFS2 schedule

Table 1.2 2012 Mandate for Renewable Fuels (billion gallons)	
Biomass-based diesel	1.000
Cellulosic biofuels	0.009
Total advanced biofuels	2.000
Cornstarch-derived ethanol	13.200
Total	15.200

inherently recognized that there would be little growth in this segment after 2015 when the minimum amount blended into gasoline would be 15 billion gallons. Growth is encouraged in so-called Advanced Biofuels whose minimum required amount in gasoline must be 21 billion gallons by 2022. The minimum total amount of renewable fuels in that year would be 36 billion gallons. EPA has the right to change the amounts for each category annually based upon its review of the fuels market of the preceding year.

The category of Advanced Biofuels refers to renewable fuels other than ethanol from cornstarch. This category includes ethanol from sugarcane, nonedible cellulose, and production from agricultural wastes such as corn stalks and bagasse. Also included in this category is biobutanol and biodiesel. A detailed breakdown of Advanced Biofuels for 2012 appears in Table 1.2.

The 2010 congressional mandate for cellulosic ethanol was off to a rough start. The mandate called for a minimum of 100 million gallons of cellulose-derived ethanol to be produced in 2010 followed by 250 million gallons in 2011 and then 500 million gallons in 2012. There were no commercial cellulose-based ethanol plants in operation in 2007 when the mandate was established nor were there any in 2011 and 2012. Only a small number of pilot and demonstration plants were in existence in those years. Congressional wisdom held that if it was mandated, the plants would appear. They didn't. Clearly, technology development cannot be mandated by legislation. A very few such plants were finally in operation in 2012. This is discussed more fully in Chapter 4 (Cellulosics).

1.2 PRESENT

As noted, EPA is entitled to change the mandates, and in 2011 they did for the following year. In 2012, the mandate for cellulosic ethanol was reduced from 500 million gallons to <12 million. Nevertheless, for failing to comply with a mandate for a virtually nonexistent product, EPA fined oil companies $10 million for waiver credits [2]. Since the purchase of such credits is essentially a fine, the costs became an invisible tax paid by the users at the gas pump. The American Petroleum Institute (API) filed a law suit against EPA in July 2012. At year's end, the Court of Appeals for the District of Columbia sided with API and declared that EPA's cellulosic quota was based on wishful thinking rather than on a realistic estimate of what the industry could achieve.

Following the decision of the DC Court of Appeals, EPA, in February 2013, issued a revised schedule for Advanced Biofuels. The revision called for a reduction in the amount of cellulosic biofuels required in 2013 from 1 billion to only 0.014 billion gallons (Table 1.3).

An intrinsic feature of the RFS2 is the Renewable Identification Number (RIN). This is a 38 digit number that must be assigned to each gallon of renewable fuel that is imported or produced in the United States [1]. It is a concept that only a bureaucrat could love. The producer or importer of the fuel assigns an RIN to each gallon, and the number undergoes changes each time the fuel changes hands or is blended until it is ultimately burned. EPA does not assign the numbers although assignment is in accordance with guidelines that they issue. The numbers indicate the calendar year of production or importation, the identity of the company, and the renewable fuel category, etc. The purpose of the RINs is to demonstrate compliance with the RFS. The RIN is *attached* to, sold, or blended into a conventional fuel. Once such blending or sale has occurred, the RIN can be

Table 1.3 Revised 2013 Mandate for Advanced Biofuels (billion gallons)			
	2012	2013	
		Original	Revised
Biomass-based diesel	1.000	1.000	1.280
Cellulosic biofuels	0.009	1.000	0.014
Total advanced biofuels[a]	2.000	2.750	2.750
Cornstarch-based ethanol	13.200	13.800	13.800
Total renewable fuels	15.200	16.500	16.500
[a]Biomass-based diesel and cellulosic biofuels.			

Table 1.4 2013 US Renewable Fuel Obligations (RVO)	
	Percent of Total Demand
Advanced biofuels	1.6
Biomass diesel	1.12
Cellulose-based fuels	0.008
Total reformulated fuels	9.03

detached, bought, or sold like any commodity. At the end of each year, renewable fuels suppliers must determine their fuel obligations (RVO) by applying the percentages in Table 1.4 to the total expected gasoline and diesel demand for the prior year.

The RVO for a supplier is the basis for the total number and type of RIN that they must submit to EPA. A supplier may sell whatever excess RINs they may have to others who are short.

The RINs can be bought and sold so that the mandates are binding or not nationwide [3]. Although it is unnecessary for each blender to use at least their share of the mandated amounts, they are obligated to buy the biofuels themselves or to buy RINs from other fuel blenders who use extra biofuels and sell extra RINs.

The RIN prices are affected by

- the gap between the price that blenders pay to buy the biofuel and its implicit price in the blended fuel;
- the transaction cost of blended RINs;
- speculation about whether the mandates will be binding in the near future.

The price gap will be positive if the mandate is binding such that blenders are forced to use more biofuels than they would have. In that instance, each gallon of fuel is either blended at a loss or blenders must buy extra RINs from their competitors. Blenders will bid the price up until it is at least as large as the loss on the marginal gallon.

In the rule establishing the RFS2, EPA also created an EPA Modern Federated Transaction System (EMTS). This is a screening system for RINs. Under this system, those purchasing or receiving RINs must certify their validity and they are responsible for any fraudulent RINs that they pass on to other buyers or submit to EPA for

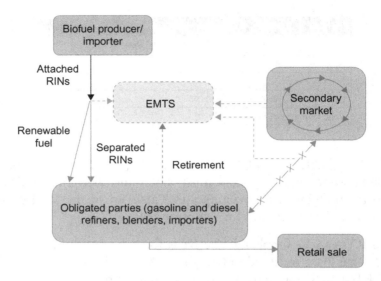

Figure 1.1 Simplified schematic of RIN Trading System. Congressional Research Service, R. Radhakrishnan, Thompson Reuters, September 25, 2012.

compliance. This is illustrated in Figure 1.1. In this figure, the solid lines indicate RINs attached to actual renewable fuels or those separated from them that may be traded among all market participants. Broken lines indicate end of year submission by obligated parties to meet RFS2 mandates. The lines denoted by x from the secondary market are for fuels separated from their RINs. All transactions from the secondary market must be cleared through EMTS. In many instances, the RINs are detached from the actual fuel at the point of initial sale or transfer and therefore may be detached from a blended fuel that has not yet been blended or sold.

Although the intent of the RINs is to monitor and ensure that US firms abide by RFS2, there have been instances of attempted fraud. In addition, there have been reports [2] that speculators have bought RINs in order to drive up their prices. While not illegal, big banks and other financial institutions supposedly amassed millions of RINs just as refiners were seeking more of them. Such activities could drive up the price of ethanol to the consumer. The value of RINs was 7 cents/gallon in January 2013, peaked at $1.43/gallon in July 2013, and fell to 60 cents/gallon in September. The banks assert they have done nothing illegal, and EPA says they have seen no evidence of improper trading. RIN price escalations, such as this, are ultimately felt by the consumer at the gas pump.

1.3 FUTURE

At best, technology forecasting is a highly uncertain art with or without the benefit of a crystal ball. The overall outlook for US energy supply and demand changed markedly only a few short years after enactment of the RFS2 in 2007. The change was driven by the advent of hydraulic fracturing (fracking) of shale formations containing oil and gas. Such formations are plentiful in the United States. In this technique, water which is mixed with chemicals and sand, is driven into fossil fuel bearing shale under high pressure. The first commercial application of fracking took place in the 1990s and by 2010 about 60% of all new wells worldwide were employing fracking [15]. Shale containing oil and gas formations in the United States and in 41 other countries currently account for 10% of global crude oil resources and 32% of those for natural gas [16].

Because of such rapid growth in production from shale, the Department of Energy (DOE) projects that the United States will be the world's largest producer of oil and gas by year end 2013, by outdistancing Saudi Arabia and Russia [17]. DOE further projects that crude oil prices will advance to $128 per barrel by 2035 along with a 16% increase in domestic consumption. This is a remarkable forecast when viewed against the notoriously poor historic 5- to 10-year forecasts of such domestic refiners. DOE views the United States as overtaking Russia and Saudi Arabia as the world's biggest oil producer as it taps rock and shale layers in North Dakota and Texas. Overtaking Saudi Arabia would occur, in their view, by 2020. A major segment of this growth is the Bakken region of North Dakota which has enabled the state to rank second only to Texas in natural gas and oil production. This region has seen a remarkable increase [18] in production of oil from about 200,000 to 1 million barrels per day in the three short years from 2010 to 2013. The DOE outlook for 2040 is that domestic US crude production could reach a phenomenal 10 million barrels per day, and that the United States would be energy independent [21] as early as 2035. In that time period, global consumption for fuel is projected to reach 820 quadrillion BTUs of which renewables will account for 110 quadrillion BTUs or 13% of the total (Figure 1.2). Accordingly, despite their rosy outlook for crude oil and natural gas, DOE nonetheless projects a US growth rate of 2.5% annually for renewable fuels. BP [20] has an even more aggressive outlook for renewables at 6.1% annual growth through 2030.

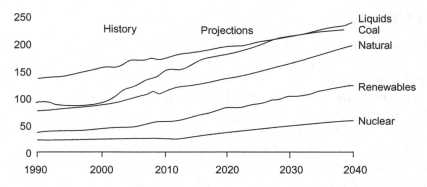

Figure 1.2 World energy consumption by fuel. U.S. Energy Administration, International Energy Outlook, 2013.

The glowing growth rate projected for renewable fuels should be tempered by the fact that as recently as 2007 fracking wasn't even anticipated as a major source of fuels production. Although renewable fuels will undoubtedly remain part of the energy picture from this time forward, the key questions are at what rate growth will actually occur and what types of renewable fuels will find a meaningful portion of that growth. Renewable fuels not only offer an opportunity for energy security but also an opportunity to reduce greenhouse gases.

Renewable fuel processes for motor vehicles must not only compete with one another but also with natural gas, hydrogen, and electric power. The following chapters address these issues.

CHAPTER 2

Ethanol by Classical Fermentation: United States and Brazil

2.1 US CORNSTARCH

The enactment of the 2005–2007 RFS by the US Congress focused on the classical production of ethanol by fermentation of cornstarch. The selection of this technology was driven by its long commercial history in the United States, and the fact that it was a readily renewable resource that would reduce dependence on fossil fuels that were seen to be in increasingly short supply. An additional factor was that it sizably produced less greenhouse gases than fossil fuels when burned. The prevailing view that fermentation of corn would lead to energy savings was challenged almost immediately by Pimental and Patzek of Cornell University [4]. They contend that such savings are illusory and that fermentation of cornstarch to ethanol would in fact require 29% more energy than would be contained in the ethanol so produced. Their study factored in total energy costs such as labor, azeotropic distillation, equipment, and fertilizers. The validity of their conclusions is the subject of some debate by such parties as H. Shapouri and J.A. Duffield at the U.S. Department of Agriculture (USDA) and M. Wang at Argonne National Laboratory [5]. The latter argue that the critique by Pimental and Patzek failed to include the value of the substantial byproducts that are produced in cornstarch fermentation.

Any uncertainty relating to ethanol and its energy savings is countered by its efficacy in lowering greenhouse gas emissions. Blended gasoline that contains 10% ethanol produces 17% less carbon monoxide and 4% less carbon dioxide than regular gasoline. Such blended ethanol also has a higher octane number (100 vs. 87) than regular unleaded gasoline and offers about the same road mileage. A more comprehensive evaluation of greenhouse gas emissions from corn ethanol by type of energy used in its production appears in Figure 2.1. This evaluation, conducted by Argonne National Laboratory, addresses raw material extraction, processing, distribution, and disposal or recycling, etc.

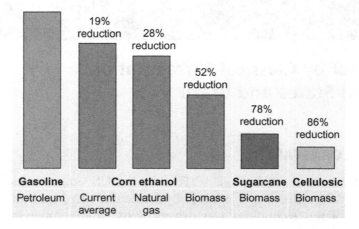

Figure 2.1 Greenhouse gas emissions of transportation fuels. Life Cycle Energy and Greenhouse Gas Emission Impacts of Different Corn Ethanol Plant Types (2007) and DOE Bioenergy Technologies Office.

The dry milling process is commonly used in the fermentation of cornstarch to ethanol (Figure 2.2). In this technique, the more profitable plants recover distillers dried grains (DDGs) as an animal feed.

In the process, cornstarch kernels are first separated from the rest of the corn plant and then ground up, The starch content of the ground up material accounts for 75% of the total corn plant. The overall process consists of four principal steps.

1. Starch hydration
2. Gelatinization of the starch
3. Enzymatic hydrolysis to fermentable carbohydrates, primarily glucose
4. Conversion of the glucose to ethanol by fermentation with yeast.

By dry milling the grain in roller mills, the finely divided starch is readily hydrated when dispersed in water. Typical grinding of the corn enables 76% to pass through a 20 mesh screen. Finely pulverized ground material reduces the downstream cooking time.

The gelatinization of the starch is necessary before it can be hydrolyzed, and the effectiveness of this step is governed by the species, its

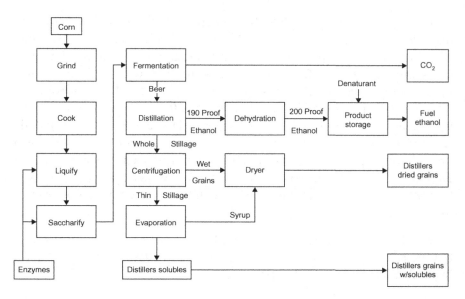

Figure 2.2 Ethanol dry milling process. American Coalition for Ethanol. © Copyright Pavilion Technologies, 2007.

time−temperature relationship, its particle size, and the concentration of the mash. Cornstarch is 80−85% gelatinized at 70−75°C, and the remainder is gelatinized as the temperature increases further to 180°C. The ground corn is combined with water and recycled stillage in a slurry tank to form a mash. The latter is continuously pumped to a steam jet heater where the mixture is maintained at 180°C and then passed through a pipe reactor designed for a retention time of 5 min. The cooked mash is first cooled to 110°C and then to 63°C.

The starch is hydrolyzed to its glucose components by a fungal enzyme (usually *Saccharomyces cerevisiae*). Approximately 80% of the total starch is converted to maltose and the remainder to branched fragments commonly called residual or limit dextrins. The dextrins are subsequently hydrolyzed to maltose in the course of fermentation. The fungal enzyme is introduced as a solution downstream of a mash cooler, and at a temperature of 63°C, and 2 min residence time, the starch is converted to a maltose/dextrose mixture.

Fermentation is dependent on time, temperature, interfacial contact, concentration, the pH of the system, and the strain of enzyme employed.

A fermentation time of 40—60 h is set by the slow conversion to maltose at a temperature <32°. Monosaccharides are faster reacting. The fermented system contains 6.5—8.5% ethanol by volume. The products are held in the fermentation tank for about 60 h during which the ethanol concentration is usually about 8.5—10%. The ethanol solution is then sent to a still from which a distillate of 50% ethanol, aldehydes, fusel oil, and waste is obtained. The condensate, frequently referred to as *high wines* is sent to a rectification column from which low boiling impurities, such as aldehydes, are removed. Since only about 10% ethanol is produced in fermentation, excess water is removed by azeotropic distillation alone or in combination with semipermeable membranes.

Fusel oil, a mixture of lower alcohols such as propanols and butanols, is removed during the course of distillation and sold as a byproduct credit against the process.

The wet milling process enables the manufacturer to separate the starch from the gluten. In that system, the corn kernels are soaked in hot water with sulfur dioxide to loosen the kernels. The grain is ground gently to loosen the germ from the kernel and is then separated. The oil, which is then recovered from the germ, is either expressed or extracted with a solvent and is used for animal feed. The degermed kernel is subsequently ground and washed to remove the hull. The remainder of the material is centrifuged to recover the cornstarch from the gluten.

The final recovered product is 95% aqueous ethanol. Automotive fuel requires 99.7% ethanol. During fermentation, each six-carbon glucose molecule is split into two three-carbon molecules of pyruvic acid that are subsequently metabolized to two-carbon molecules of ethanol as shown in Figure 2.3. NAD represents adenine dinucleotide and is an enzyme cofactor which successively donates and removes hydrogen in a series of cyclic redox reactions. NADH represents its reduced form in which it is bonded to the hydrogen and NAD^+ is its transient oxidized version in which it has given up its hydrogen.

Production of US fuel grade ethanol soared [6] from 2.1 billion gallons in 2002 to 13.9 billion gallons in 2011, a nearly sevenfold increase. The number of plants increased from 100 to 123 in this time period. The rapid growth was nearly entirely driven by fermentation of cornstarch

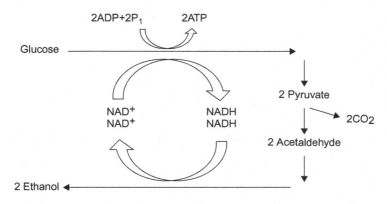

Figure 2.3 Glucose conversion to ethanol.

and the mandated RFS. It was secondarily aided by duties on imported ethanol and a tax incentive of $0.45 per gallon of nearly pure ethanol blended with gasoline. The tax incentive had first to be taken as a credit against the blender's tax liability, and any excess over this could be claimed as a direct payment from the Internal Revenue Service. Further impetus to the growth in cornstarch ethanol was the parallel phase out of t-butylmethyl ether as an octane improver.

Since 1 acre of corn is required to produce 328 gallons of ethanol [7] production of motor fuel ethanol quickly consumed 40% of the corn produced in the United States. Consumption for fuel had been only 14% of the total crop in 2005.

As would be expected, the competing demands for food and fuel drove corn prices from $2.78 per bushel in 2006 to $7.25 per bushel in mid-2012. A significant contributing factor was the severe drought that struck the corn producing mid-western part of the United States in 2012. As a consequence, profitability in corn production fell sharply for simple plants in which corn oil was not recovered from the DDGs as a byproduct credit. Beginning in the summer of 2012, the prices for ethanol and corn reached levels where the production costs at such relatively simple ethanol plants exceeded revenue [9]. Figure 2.4 provides a graphic illustration of the economics affecting plant shutdowns in 2012 and 2013. As would be expected, there were very few announcements for new cornstarch-based plants. Indeed, by January 2013 the number of idled corn-based ethanol plants had grown to at least 20. Those plants that recover corn oil have gross margins of at least

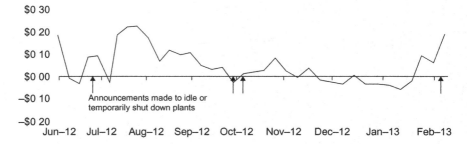

Figure 2.4 Estimated average margins of ethanol plants without corn oil recovery (2012–2013). U.S. Energy Information Administration and USDA Agricultural Marketing Service.

15–20 cents/gallon greater than those plants that do not. Nearly 10% of the US ethanol plants ceased production because the drought and food competition pushed ethanol prices too high [8].

In establishing the RFS, Congress had assumed that gasoline consumption would steadily increase. Instead, it has declined from 14.2 billion gallons per year in 2007 to 13.0 billion gallons in 2012. The reasons for this are not clearly understood. A change in driving habits and increased mileage for new automobiles are frequently cited as primary factors. In view of all this, the $0.45 per gallon Federal tax incentive on ethanol was allowed to expire at the end of 2011.

The estimated cost of production for a hypothetical and relatively simple plant that does not recover corn oil is presented in Table 2.1 for illustrative purposes. These economics are for a 49.5 million gallon average size plant in the US Midwest in the third quarter of 2013. With corn priced at $4.53 per bushel, the variable cost of production is $2.27 per gallon. With DDGs as a byproduct at $2.28 per ton, the net ethanol cost of production declines to $1.82 per gallon. The net cost of ethanol from a plant that recovers corn oil is about $0.17 per gallon less, i.e., $1.65 per gallon. With an arbitrary 15% return on total investment of $190.7 million, the cost of the ethanol becomes $2.39 per gallon. This is substantially greater than the futures price at the Chicago Board of Trade of about $1.79 per gallon in late 2013 and should be viewed against the average US price for gasoline in 2013 of $3.50 per gallon.

An additional factor that hinders growth in cornstarch-derived ethanol is the so-called blending wall. This limits the ethanol content in US

Table 2.1 Fermentation Ethanol from Cornstarch 49.5 Million Gallon Plant (2013)	
Total Fixed Investment	$190.7 million
Cost of Production	cents/gallon
Raw materials	
Corn (0.383 bu/gal @ $4.53/btu)	173.50
Chemicals and enzymes	10.00
Utilities	
Electrical power ($1.76 kwh/gal @ 7 cents/kwh)	12.32
Natural gas (11,900 btu/gal @ $3.50/$10^6$ btu)	4.17
Steam, lp (24.6 lbs/gal @ $10.93/$10^3$ lbs)	26.88
Subtotal (total variable costs)	*226.87*
Total fixed costs	*40.16*
Total ex byproduct credit	**267.03**
Byproduct credit (DDGs; 3750 tons/gal @ $228/ton) (85.50)	
Net cost of production	**181.53**

gasoline to 10% and is referred to as E-10. An effort to raise this limitation to 15% (E-15) has been strongly resisted by refiners, automobile manufacturers, and the American Automobile Association [10,11]. The arguments against E-15 are that greater alcohol content in gasoline would accelerate engine wear and failure, and would otherwise damage the fuel system. In a lawsuit brought by the API [12] against EPA, the US Court of Appeals for the District of Columbia decided in favor of EPA in allowing the sale of E-15 in vehicles built in 2001 and later. The API subsequently filed a brief with the Supreme Court in which they requested that the decision of the DC Court of Appeals be overturned. This was finally resolved in August 2013 when the Supreme Court decided in favor of EPA. In the meantime, the blend wall has impeded ethanol production since the domestic market for it is saturated and the Supreme Court's decision has had little immediate impact on greater addition of ethanol to gasoline. The vast majority of US service stations don't offer E-15 except for about two dozen in the US Midwest. Until US refiners begin to supply greater amounts of E-15 at their service stations, and consumers agree to purchase it, the options for refiners are to decrease production or to export the surplus. At year end 2013, EPA further complicated the situation by proposing a reduction in the mandatory requirements for ethanol and other renewable fuels in 2014. EPA's proposal has met with furious

Table 2.2 EPA Proposed RFS Standards for 2014	
Cellulosic biofuels	0.010%
Biomass-based diesel	1.18%
Advanced biofuel	1.33%
Total renewable fuels	9.20%
The percentages represent the ratio of renewable fuel volumes to total nonrenewable gasoline and diesel volumes.	

resistance by the renewable fuels industry and growers in the corn producing states and is illustrated in Table 2.2. The blend wall controversy was not fully resolved going into 2014.

The uncertainty that surrounds the future of cornstarch-derived ethanol has been reflected in the decision of several firms to divest their equity in cornstarch ethanol plants. Bunge North America, for example, decided in late 2013 to sell its equity in a 54 million gallon plant in Vicksburg, MS. In addition, in early 2014, GTL Resources and Illinois River Energy decided to sell their 125 million gallon plant in Rochelle, IL. The plant is relatively new and started up in 2006 and is capable of producing ethanol from sugar cane and sorghum in addition to cornstarch.

A listing of US ethanol producers is presented in the Appendix.

2.2 BRAZILIAN SUGAR CANE

After the United States, Brazil is the largest producer of fuel ethanol, and together the two countries comprise about 88% of global production. At the beginning of 2012, Brazil produced about 5.6 billion gallons of ethanol and accounted for 25% of total global ethanol that was used for automotive fuel. Brazil's 40-year fuel ethanol program is regarded as a model for other countries and is arguably based on the most efficient technology for sugar cane production worldwide. It uses modern equipment, inexpensive sugar cane. The residual sugar cane waste, called bagasse, is burned to produce heat and power to produce a byproduct credit against the cost of cane sugar production. This leads to a positive overall energy balance of 8.3–10.2 as cited earlier [14]. In 2010, the U.S. EPA designated Brazilian sugar cane ethanol as an *Advanced Biofuel* because of its 61% reduction in total life cycle greenhouse gas emissions including its indirect land use change emissions.

In contrast to the United States, the Brazilian government has required ethanol to be blended into gasoline for passenger vehicles since 1976. Ethanol content in Brazilian gasoline is significantly higher than in American vehicles and averages 20–25%. Brazilian automobile producers have developed flexible fuel vehicles that can run on any blend in gasoline, usually 20–25%, and even 100%. In terms of energy equivalency, sugar cane ethanol represents as much as 18% of the country's total energy consumption for transportation.

Being more vulnerable than the United States during the 1973 OPEC oil embargo, Brazil quickly sought an end to its dependency on imported oil by developing an ethanol fuel economy based on its extensive production of sugar cane. The blending of ethanol into gasoline was Brazil's second major foray into this activity: the first approach occurred when submarine attacks by Nazi Germany during World War II seriously threatened Brazil's crude oil supplies. At the time, the country lacked a meaningful base in domestic crude oil production. The mandatory blend of ethanol became as large as 50% in 1943, and then quickly fell to nearly zero at the War's end. There was a rapid resurgence in blended ethanol with the above-cited threat by the OPEC Oil Embargo in 1974. Having been burned twice by its dependency on imported oil, Brazil, in the following year, decided to phase out gasoline derived from fossil fuels. The history of blended ethanol into Brazilian gasoline is depicted in Table 2.3.

Production of ethanol from sugar cane is more economical than cornstarch. For one thing, the initial step in which corn starch must

Table 2.3 Ethanol Blends in Brazilian Gasoline [32]	
Year	Percent
1976	11
1978	18–23
1981	12–20
1985	20
2005	22
2009	25
2011–13	25

first be degraded to a fermentable carbohydrate substrate can be eliminated [13]. In its place, cane sugar can be fermented directly to ethanol. The low cost of domestic sugar cane coupled with the efficient use of bagasse for fuel affords it at least an eightfold positive energy balance overall. This is unlike cornstarch fermentation in which the energy balance may be neutral or even negative. As cited earlier, the advantageous energy balance for using sugar cane has led EPA to characterize it as an *Advanced Renewable Fuel*. Most of the industrial processing in Brazil is accomplished in a highly integrated manner in which cane sugar production, ethanol processing, and electrical power generation are linked.

Sugar cane is washed, chopped, and shredded by revolving knives. The resulting product is extracted to produce a juice that is 10−15% sucrose and bagasse. The cane juice is filtered and concentrated by evaporation to produce a syrup of clear crystals surrounded by molasses. The latter is sterilized and then used to produce ethanol by fermentation as illustrated in Figure 2.5.

Molasses, which is recovered in the refining of sugar, is combined with aqueous ammonium sulfate and yeast at 30°C and about 4.5 pH. Fermentation time lasts from 4 to 12 h and yields a product with an

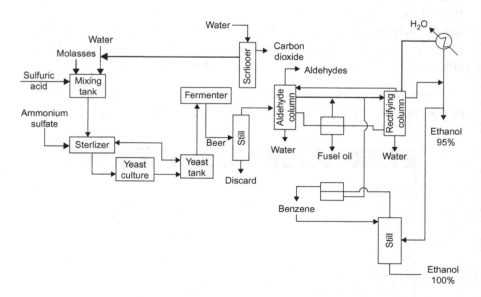

Figure 2.5 Block flow diagram: sugar cane fermentation to ethanol.

ethanol content of 7—10%. This is sent to rectification as shown. The raw molasses is clarified and the solids are returned to the cane fields as mulch and fertilized. The clarified molasses is divided into two streams. Approximately 5% of the total volume is charged to the yeast fermenters. The remaining 95% of the clarified molasses is sterilized and concentrated to lower the cost of downstream purification of the fermented product. The concentrated juice is combined with the prepared yeast and additional nutrients. The recovered 95% ethanol is subjected to azeotropic distillation and/or processed through semipermeable membranes to yield a fuel grade product of 99.7% ethanol. The molasses feedstock is comprised of 55% sugar of which 35—40% is sucrose and the balance is invert sugar (equimolar amounts of glucose and fructose). The greater simplicity in ethanol production from sugar cane compared to cornstarch is reflected in its lower fixed capital investment.

Energy use associated with sugar cane ethanol production is derived from three primary sources: (i) sugar cane growth, (ii) ethanol production, and (iii) product distribution. About 36 GJ are used to plant, maintain, and harvest 10,000 square meters of sugar cane. This includes fertilizers, fuel, and pesticides. In comparison, 3.6 GJ per 10,000 square meters are required for ethanol production while generating 155.6 GJ. The potential power generation from this activity could range from 1000 to 9000 MW by most estimates. The same 10,000 square meters requires about 2.8 GJ of energy. After taking all three categories into consideration, the net energy for sugar cane ethanol [32] is a positive ratio of 8.

In a study by the consulting firm, Chem Systems, there is a savings of slightly more than 21% in the battery limits capital cost for ethanol production from sugar cane. This is because a sugar cane process eliminates the necessity of first converting cornstarch to sugar for fermentation. Additional savings accrue to Brazilian producers' net utility costs as provided by energy savings. Brazilian producers are reportedly [33] able to produce ethanol for 22 cents per liter compared with 30 cents per liter for cornstarch-derived product.

As might be expected, Brazil is the largest exporter of ethanol with a 2007 volume of nearly 1 billion gallons. Although the United States is potentially the largest market for Brazilian ethanol, the United States imposes a duty of 54 cents/gallon to protect its own industry

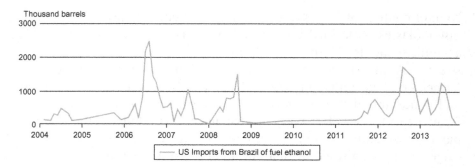

Figure 2.6 US imports from Brazil of fuel ethanol. U.S. Energy Information Administration.

and to encourage its growth. Prior to Congressional action on the Reformulated Fuels Standard in 2006, exports of Brazilian ethanol to the United States were about $1 billion. This volume fell by nearly half in 2007 with the growth of cornstarch fermentation in the United States, and had remained largely negligible through 2012 (Figure 2.6).

2.3 UNITED STATES VERSUS BRAZIL

Sugar cane cultivation requires a tropical or subtropical climate with a minimum of 24 in. of rainfall annually. Consequently, production in the United States is limited to Florida, Louisiana, Hawaii, and Texas. Nearly 86% of the US production of 26.7 million tons is shared almost equally between Florida (13.1 million tons) and Louisiana (10.1 million tons). Nonetheless, production compared to cornstarch is small with the United States ranking fifth worldwide. A comparison of the differences in the key characteristics for ethanol production between the United States and Brazil is shown in Table 2.4.

In 2007, former president George W. Bush and Brazil's president Luiz Inacio Lula da Silva sought means to promote the use and production of sugar cane ethanol throughout Central and South America. They agreed to share technology and set international standards for biofuels. The intent was to transfer Brazilian technology to such countries as Honduras, Costa Rica and to enhance trade between them and the United States. Even though the United States has imposed a 54 cent/gallon tariff on imported ethanol since 1980, these countries are exempt from this duty as long as the ethanol produced from foreign feedstocks doesn't exceed 7% of the previous year's US consumption.

Table 2.4 Comparison of Brazilian and US Ethanol Production		
	United States	Brazil
Feedstock	Cornstarch	Sugar cane
Acres used for ethanol (2006)	25 million	9 million
Gallons of ethanol/acre	321–424	727–870
Energy balance	1.3–1.6	8–10
Reduction in greenhouse gases	10–30%	86–90%
Ethanol share of gasoline market	10%	50%

Additional quotas are permitted if these countries produce at least 30% of their ethanol from local sugar cane. There is a maximum amount of 35 million gallons permitted under this quota. As a result, several countries have been importing dilute, aqueous Brazilian ethanol, dehydrating it locally, and then exporting it to the United States as anhydrous fuel ethanol. As would be expected, American farmers have protested this loophole in the tariff.

The USDA provides loans to sugar cane producers that guarantee them a minimum price regardless of true market conditions. At the end of the loan period of usually 9 months, sugar growers and processors must make one of two choices:

1. turn over to the government the sugar so produced as payment for the loan
2. sell the sugar on the market if the going price is greater than the loan.

The current rate for a loan is 18.75 cents per pound of raw cane sugar. The amount of sugar a company is permitted to sell in a given year is its share of the market as determined by USDA. If a company produces more sugar than their allotment, they are forbidden to sell it and must store the excess at their own expense.

US sugar imports are strictly controlled by tariff rate quotas (TRQ/s). These are established annually and are the amount of sugar that can enter the United States at low or zero duty. For 2012, the United States allocated a quota of 133,634 tons of raw sugar cane from Brazil. The United States also has a Sugar to Ethanol program under which excess sugar generated by generous price supports can be sold to ethanol producers at a significant loss. Under this program, ethanol producers can pay for sugar the equivalent of what they pay for less expensive corn. Since it is more profitable to convert sugar cane to sugar *per se* in the United States, there are no operating plants for the conversion of sugar cane to ethanol.

Imported ethanol derived from sugar cane is attractive as an Advanced Biofuel under the Reformulated Fuels Standard discussed in Chapter 1. This drives the relatively large volume of ethanol imported from Brazil since 2008. The volume of imports cannot be attributed to less expensive Brazilian ethanol, which on a particular day, November 29, 2012, was $2.60/gallon at US Gulf terminals compared to $2.65/gallon at the Brazilian port of Santos. If one assumes that freight costs are 20 cents per gallon, the landed cost of Brazilian ethanol is $2.85 per gallon. No US blender would therefore have purchased Brazilian ethanol based on market economics. The answer to Brazilian imports therefore lies in the RFS passed by Congress (Chapter 1). This requires a certain amount of Advanced Biofuels to be consumed annually. Since cornstarch-based ethanol has a less favorable greenhouse gas reduction rating, it can only qualify as a *renewable* biofuel and therefore cannot compete with Brazilian ethanol or other advanced biofuels to fulfill the Congressional Mandate. This means the relevant economic comparison is whether US produced biofuel or Brazilian produced sugar cane ethanol is the cheapest source for meeting obligations under RFS2.

CHAPTER 3

Ethanol from Cellulose

The attractiveness of waste cellulose for the production of ethanol is that it avoids competition between food and fuel. Cellulose is ubiquitous in that it constitutes 50% of the woody structure of most plants, 10% of the dry weight of leaves, and 98% of cotton fiber. It can be used in many forms, such as wood chips, sawgrass, miscellaneous crop residue, and last, but not least, municipal solid waste. Cellulose is a polysaccharide composed of straight chain 1,4-beta-D-glucose units with the following general structure (Figure 3.1).

Accordingly, at first glance it would appear that the fermentation of cellulose would be a ready trade-off for cornstarch. Such similarity is only superficial, however. Unlike starch whose glucose polysaccharide structure is highly branched, cellulose molecules are very linear with a substantial degree of hydrogen bonding that contributes to a compact highly crystalline structure. This tight bonding and crystallinity leads to a highly compact structure that inhibits access by enzymes. Second, cellulose is closely associated with hemicellulose that accounts for 20% of the structure of most plants. Unlike cellulose, hemicellulose is largely composed of five-carbon monosaccharides, primarily xylose, which are not readily metabolized by *Saccharomyces cerevisiae* to ethanol. In addition, cellulose is intimately associated with lignin, a complex cross-linked polymer of phenylpropane units. Lignin is the most recalcitrant segment of plant cell walls and its presence further reduces the surface area available for enzyme penetration and attack.

The combined barriers afforded by crystal structure, hemicelluloses, and the presence of lignin are the basis for various techniques to convert cellulosic materials to renewable fuels. Developers of such techniques fall into two broad groups: (i) those who employ fermentation partially or solely throughout the process and (ii) those who first convert the cellulose to a synthesis gas that can be converted either biochemically or synthetically to fuels. In the first instance, cellulose must first be "softened up" or pretreated to make it susceptible to

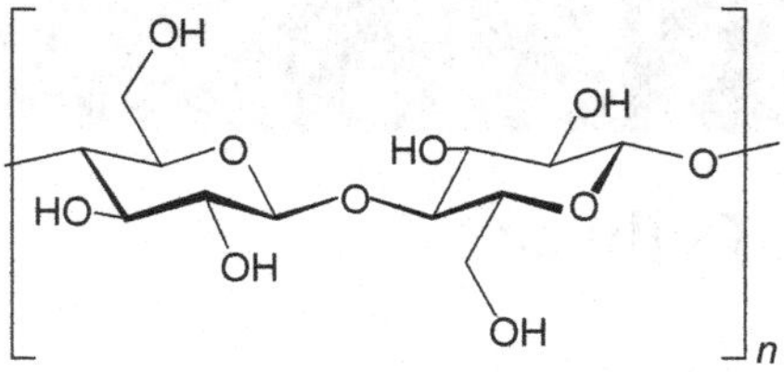

Figure 3.1 Cellulose structure.

enzyme attack. There is an important issue confronting the first group and that is the utilization of the five-carbon carbohydrates (xylose and arabinose) that are present in lignin. For example, the hydrolyzate of corn stover is 30% xylose. Because of this, the economic viability of a process is contingent upon the ability to ferment xylose and arabinose to ethanol efficiently. Engineered enzymes are under development that may be capable of this transformation [35]. Some species of bacteria, such as *Clostridium thermocellum*, are capable of direct conversion of cellulose into ethanol. However, this species also produces acetic and lactic acids and so lowers the efficiency of the process.

The synthesis gas approach destroys the cellulose structure by partial oxidation in which it converts it to CO and hydrogen. There is also another technique under current investigation in which the amount of lignin in a plant is reduced during growth but not to the degree that the plant is unable to retain its physical structure. Investigators at Ghent University in Belgium and the University of Dundee in Scotland find that manipulation of caffeoylshikamate esterase, an enzyme in lignin biosynthesis, is effective in controlling lignin formation [22]. Although four times more cellulose can be converted to glucose in such plants, the plants themselves are smaller and less sturdy than wild types. Considerably more effort is needed before this last approach can be considered commercially.

The highly ordered structure of cellulose has encouraged the development of a number of pretreatment systems. Among these are acid hydrolysis, steam explosion, and the Organosolv process. Organosolv entails the use of methanol or ethanol at 100−250°C to solubilize the lignin and to separate it from the cellulose. To be economically attractive, the process requires the recovery and recycling of the solvent alcohols.

The oldest pretreatment technique is dilute mineral acid hydrolysis. This approach continues to be investigated and uses such common acids as HCl and H_2SO_4. Its primary drawback is the formation of furfural and hydroxymethyl furfural from the glucose. Both furfural compounds are enzyme inhibitors. In addition, acid neutralization is required before enzyme-catalyzed production of glucose to ethanol can be considered. BlueFire Renewables of Irvine, CA is currently pursuing this process for conversion of cellulosic wastes to ethanol. In 2007, they received $40 million in funding from the U.S. Department of Energy to build a $19 million plant in Mississippi.

In another approach by Shell Oil, tertiary polyamide additives, such as polyvinylpyrrolidone and polyalkyloxazolines, are used to break down the cellulose into its component carbohydrates [34]. Ground up and homogenized wheat straw that is treated with these additives improves cellulose conversion to glucose by 50%.

Steam explosion is the most commonly used pretreatment process. After mechanical size reduction, the cellulose is rapidly heated to $180-240°$ with high-pressure ($1-3.5$ MPa) steam. This is followed by an explosive decompression that results in a rupture of the rigid fibers of the cellulose. The sudden pressure release defibrillates the cellulose bundles and affords greater accessibility to the enzymes for hydrolysis and fermentation. The presence of dilute nitric or sulfuric acid accelerates the process. The steam rapidly heats the cellulose to its target temperature without excessive dilution of the resulting carbohydrates [23].

The DuPont Company has introduced an enzyme system entitled Accellerase that converts pretreated cellulose into fermentable C_5 and C_6 carbohydrates simultaneously. Its first commercial plant for this technology, which is also apparently for license, is being built in Nevada, IA. Optimum reaction conditions comprise 0.22 ml/g of cellulose at pH 5.0 and 50°C. In 3 days time, glucan and xylan conversions are about 90% and 80%, respectively. Performance drops precipitously above pH 5.5 and 50°C. DuPont reports that Accellerase is effective in systems that combine cellulose hydrolysis and fermentation of the resulting five- and six-carbon carbohydrates.

Among the more notable companies who use pretreatment followed by liquid phase enzyme action are Raizen, Beta Renewables, DuPont/ Betamax, and Dong Energy.

Raizen is a $12 billion joint venture between Royal Dutch Shell and Cosan, Inc., which is the largest energy company in Brazil and the fifth largest company there overall. Raizen plans to bring a 10 million gallon ($90 million) bagasse-based ethanol plant onstream at Piracicaba, Sao Paulo by the end of 2014. The plant will be integrated with Raizen's sugarcane-based ethanol unit at the same location. The proposed plant will use Iogen's steam explosion technology for pretreatment coupled with cellulase enzyme fermentation provided by Novozymes. The Iogen steam explosion segment of the process exposes the bagasse to pressures of 200–450 psig for <10 min after which the reactor is rapidly depressurized to 1 atm. A similar steam explosion system has been developed by SunOpta of Canada.

Iogen Corp. is a closely held company based in Ottawa, Ontario. In 1974, Iogen Corp. formed Iogen Energy jointly with Royal Dutch Shell to develop technology for the production of ethanol from cellulose. They claim to have built the world's first demonstration scale facility to convert cellulose to ethanol via enzyme catalysis. The facility was designed to process 40 tons/day of wheat straw using enzymes produced in an adjacent enzyme production plant, and they claim that it could produce >340 l of ethanol per ton of cellulose (wheat straw) fiber. On April 30, 2012, Shell withdrew from its Iogen joint venture and sold its stake in the company to its other joint venture with Raizen. Iogen also had a Bio-Products business that was sold to Novozymes (see above) in early 2013. The Bio-Products business made and sold enzymes used for pulp, paper, and textile processing. Iogen will design and engineer the front end of the Raizen plant. Novozymes is an industrial enzyme developer and supplier and is located in Bagsvaerd, Denmark. They are the world's largest enzyme company. As a consequence of all these liaisons, the proposed Raizen plant directly and indirectly can be considered as based on Iogen technology. At year end 2013, Iogen had 26 patents issued and in various stages of prosecution. An engineered enzyme from the fungus *Trichoderma reesei* is employed in the hydrolysis step.

An interesting development that combines pretreatment with fermentation is the Proesa process introduced by Beta Renewables, a joint venture company between TPG (an independent investment company) and a sister company (Chemtex Agro). The Proesa process is illustrated in Figure 3.2 where hydrolysis and fermentation proceed simultaneously.

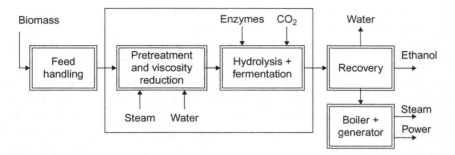

Figure 3.2 Proesa process. Chemtex Group; Presentation at IEA Conference Vienna, November 14, 2012.

The Proesa technology may eventually be protected by at least 14 US patents. Of this 14, 7 were published by the beginning of 2014 and are summarized as follows:

1. 20140034253 Process for Acetic Acid Removal from Pretreated Biomass
2. 20130313472 Pretreated Biomass Having Enhanced Enzyme Accessibility
3. 20130168602 Presoaking Process for Biomass Conversion
4. 20130158253 Process for Recovering Sugars from a Pretreatment Stream of Lignocellulosic Biomass
5. 20130149761 Methods to Recover Sugars of Pretreated Biomass Liquids
6. 20130146049 Process for Recovering Sugars from a Pretreatment Stream of Lignocellulosic Biomass
7. 20120211427 Regenerative Purification of a Pretreated Biomass Stream.

The development is the result of more than 7 years of research at a total cost of about $205 million.

The Proesa Process was developed by Biochemtex Laboratories of Rivalta Scrivia, Italy in a partnership with Beta Renewables. This system is similar to one under development by Mascoma Corp. of Lebanon, NH, which uses genetically modified cellulases. The 20 million gallon Beta Renewables ethanol plant in Crescentino, Italy, started up at the end of 2013 and is reportedly the world's largest. The new plant operates on wheat straw and products from the cultivation of a Mediterranean species of reed. As in the case of Raizen, Novozymes' enzymes are also used and in this instance, Novozymes has a 10% ownership on the Beta

Renewables plant. The Proesa system has faster reaction times than conventional fermentation, possibly because of higher reaction temperatures. Variability in reaction conditions enables flexibility in C_5 carbohydrate production. Beta Renewables reports that its Proesa Process can recover one unit of ethanol for every 4–5 units of biomass. Translating this into economics, they claim that their projected cost for cellulose feedstock is $40–50 per metric ton or about $80–100 per metric ton of recoverable carbohydrates and $150 per ton for the enzymes used in the system. This is equivalent to $230–250 per metric ton for the contained sugars. In terms of ethanol production, this amounts to $50 per metric ton for the cellulose feedstock plus $150 for the enzymes and another $50 for fixed and variable operating costs. The total cost of production is an attractive $370–440 per metric ton or $1.11–1.31 per gallon. To this must be added a return on capital for the new plant and marketing and distribution costs. It is assumed that depreciation is included in their fixed costs Reportedly, the capital investment for the plant was 90 million euros or $123 million at 1.37 euros per dollar in March 2014.

Novozymes is engaged as an enzyme supplier to more than half of the enzyme-based cellulosic projects as illustrated in Table 3.1.

A process similar to that of Beta Renewables is the one developed by Clariant, a Swiss company, and is described by them as the Sunliquid process [37,38]. In their system, cellulose particles of <2 mm in size are obtained by pretreatment and then processed to ethanol via

Table 3.1 Cellulosic Enzyme Suppliers

Company	Million Gals.	Start-Up	Supplier
Abengoa	25	2014	Dyadic
Cofco-Sinopec	15	2014[a]	Novozymes
DONG	15	2013	Novozymes/DuPont/DSM
DuPont	28	2014	duPont
Fiberight	6	2013	Novozymes
Graalbio	22	2013	Novozymes
Mascoma	20	2013	Generated *in situ*
Mossi & Ghisolfi	13	2012	Novozymes
Mossi & Ghisolfi	20	2014	Novozymes
Poet	25	2013	DSM
Shenguan	6	2013[a]	Novozymes
[a]China.			

enzymes produced on-site. The enzyme production is integrated with lignocellulose fermentation in which both five- and six-carbon carbohydrates are converted to ethanol in "one pot." Clariant is located in Muttenz, Switzerland, and is essentially a specialty chemical company that is an outgrowth of acquisitions from Farberbe Hoechst, Sud-Chemie, and British Tar Products. In January 2014, they joined forces with Mercedes-Benz and Halterman to demonstrate the effectiveness of Sunliquid20® (a 20% ethanol blend) as a high-performance fuel. The role played by Halterman is to blend the ethanol into gasoline.

In another variation of cellulose processing, Dansk Olie og Naturgas (DONG Energy) first cuts the cellulose into small pieces and then heats them under pressure to open the lignin structure to enzyme-catalyzed hydrolysis. Enzymes are then added to the cellulose–lignin mixture to convert part of the cellulose to lower carbohydrates. The resulting low viscosity liquid is then sent to a traditional fermenter to produce ethanol and carbon dioxide. Water is removed from the product by a molecular sieve to yield 100% ethanol. DONG's subsidiary, Inbicon, has operated a demonstration unit (termed a biorefinery) at Kalundborg that is capable of producing 5.4 million liters of ethanol annually from 30,000 tons of straw. Lignin pellets (13,000 tons) and C_5 molasses (11,100 tons) are coproduced. The biorefinery is integrated with the adjacent Asnaes power station (also owned by Dong). The power station sends waste steam to the biomass refinery where it is used for cellulose pretreatment. They claim that the lignin process is so clean that no further treatment is necessary when it is used in the power plant. The biomass refinery can produce enough thermal and electrical energy to offset up to 50% of ethanol plant's utilities costs. The new biomass refinery is the focal point of an Inbicon Biomass Technology Campus.

Paris-based Deinove's process also accomplishes cellulose processing and fuel production simultaneously. The company uses Deinococcus bacteria without additives. The bacteria are stable under operating conditions of 40–60°C. An intriguing feature is the suggestion that the technology can be used in existing ethanol plants without the need for major new investments. If correct, this suggests that existing cornstarch facilities can be used for cellulose processing. The Deinove's system was in the laboratory stage of development during 2012–13 with plans to move next to the pilot plant stage.

Colorado-based Zeachem [24] employs a hybrid combination of biochemical and thermochemical processing that reportedly preserves the best of both in terms of yield and efficiency. Supposedly, no new microorganisms are employed. After chemical fractionation to separate the lignin, the hemicellulose and cellulose are fermented with aceto-genic microbes to produce acetic acid. Since no carbon dioxide is produced, carbon efficiency is close to 100%. The acetic acid is subsequently esterified with ethanol to produce ethyl acetate, and this is then hydrogenated to yield additional ethanol. The hydrogen for the process is recovered from synthesis gas obtained from the synthesis gas produced by gasification of the lignin recovered from the fractionation in the initial step of the process. The remainder of the synthesis gas is burned to produce the steam and power to drive the process (Figure 3.3). The net effect is that two-thirds of the energy in the ethanol product comes from the C_5/C_6 carbohydrates and one-third from the lignin. In a sense, everything is used from the pig except the squeal. A process option to send the synthesis gas produced by gasification of the lignin to a fermenter where it can be converted directly to ethanol is discussed in the next chapter.

An entirely different approach to those cited above is that of Sweetwater Energy. Sweetwater, located in Rochester, NY, was founded as Sweetwater Ethanol in 2006 with the intention of enabling farmers to produce ethanol from crops right on their own farms. The company underwent a name change in 2009 and focused on new cellulosic sugar technology and built a pilot plant for the extraction of aqueous sugar from silage. The undisclosed technology is the subject

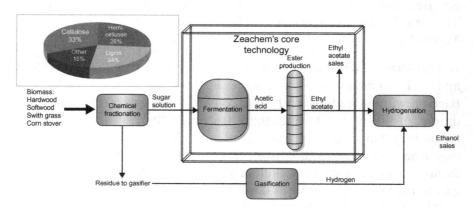

Figure 3.3 Zeachem Process. Zeachem, Inc., 165 South Union Blvd, Lakewood, CO.

of four US Patent applications from 2010 to early 2014. The intent is to sell the recovered aqueous solutions of five- and six-carbon sugars to biorefiners for production of ethanol and other applications.

Traditional yeast fermentation of cornstarch yields one molecule of carbon dioxide for every molecule of ethanol. Consequently, the carbon efficiency for the Zeachem process is nearly 100% compared to 67% for yeast fermentation. Potentially, plant yields of 1.35 gallons of ethanol per ton of dry cellulose are possible. Zeachem does not specifically disclose its technique for the separation of cellulose and hemicellulose from lignin. Related efforts by others [25] show that this can be readily done by solvent fractionation with methylisobutyl ketone/ethanol (MIBK) at about 140° in the presence of sulfuric acid. Phase separation occurs upon the addition of water. Lignin is precipitated and the cellulose is retained in the MIBK phase while the hemicellulose is in the aqueous phase. More recently, it was reported [26] that ionic solvents, such as 1-butyl-3-methylimidazoline, are quite effective in the separation of cellulose, hemicellulose, and lignin from one another.

One firm, KIOR of Pasadena, TX, has decided to finesse cellulose pretreatment entirely by converting cellulose directly to fuels. The brute force approach passes waste cellulose through a fluid bed catalytic cracker (FCC) of the type commonly used for crude oil in refinery operations. Their $190 million unit in Columbus, MS, is designed to produce 11 million gallons of fuels per year. The ratio of gasoline/diesel/fuel oil is expected to be 35/40/25 which is substantially the same as is obtained from conventional crude oil processing. A second unit is planned for Natchez, MS, in 2014. As is the case of most new technology, there have been some operating problems at the Columbus plant where the projected full year production for 2013 was only about 920,000 gallons. KIOR's approach does not offer individual renewable fuels but a renewable feedstock from which conventional fuels can be produced. In March 2014, continuing operating and financial difficulties led the company to express doubts about its ability to continue as a viable firm. In a 10-K Report, they announced the likelihood of near term bankruptcy by April 2014. If so, there is the possibility of abandonment by KIOR to their approach to renewable fuels. KIOR might be the latest instance in the short history of renewable fuels development where a number of firms defaulted on state, federal, and private funding. In some instances the technology was taken over by a third party, who with their modifications and resources, proved to be successful.

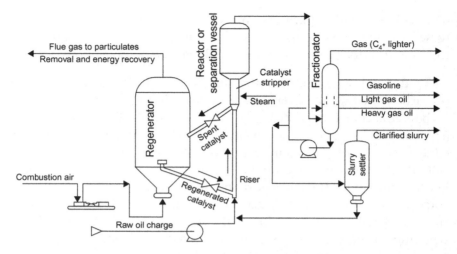

Figure 3.4 Diagram of the fluid catalytic cracking process. U.S. Energy Information Administration, modified from OSHA Technical manual.

In early 2014, West Texas Intermediate crude oil was priced relatively cheaply at $93.67 per barrel which is equivalent to $16.15 per million BTUs. Wood waste in comparison had a value of only $4 per million BTUs at that time. This is a significant advantage for waste cellulose provided that its processing does not have inordinate and capital investment costs. The configuration for a **KIOR FCC** plant (Figure 3.4) appears to be largely the same as a conventional FCC unit where fluidized hot catalyst from a regenerator flows into a riser tube reactor. The incoming cellulose and recycled slurry are vaporized and cracked when they contact the hot catalyst. Coke forms on the catalyst and is burned off. The crude oil feedstock in a conventional FCC process has an initial boiling point of about 340°C and an average molecular weight of 200−600. Waste cellulose falls significantly outside these ranges. Barring operating problems because of its uniqueness, the **KIOR** process could offer a significant advantage over fossil fuels.

Some species of bacteria, such as *Clostridium thermocellum* and *C. thermosaccharolyticum*, can convert cellulose directly to ethanol. This organism employs a complex cellulosome to convert the cellulose and hemicellulose, but it also produces additional products such as lactic and acetic acids, thereby lowering the overall yield of ethanol.

CHAPTER 4

Synthesis Gas-Based Fuels

A growing number of parties have elected to abandon cellulose pre-treatment/fermentation (see Chapter 3) in favor of cellulose conversion to synthesis gas as a way to overcome the intractable nature of cellulose to processing. In so doing, they are lifting a page from the classical gasification of fossil fuels such as coal in its conversion to liquid fuels. The synthesis gas (a mixture of carbon monoxide and hydrogen) is subjected to either partial oxidation (a) or steam reforming (b) as shown by the following general reactions using methane as an example (Figure 4.1).

The ratio of carbon monoxide to hydrogen can be readily adjusted by using the water gas shift reaction (c) in a second step (Figure 4.2).

Once the synthesis gas has been cleaned of any deleterious components, it makes little difference whether its conversion to liquid fuels proceeds enzymatically or chemically. Chemical conversion to liquid fuels can take place either by the well-established Fischer–Tropsch (FT) or ExxonMobil methanol-to-gasoline (MTG) systems or by newer enzyme conversion processes that use such acetogens as *Clostridium ljungdahlii* [27] and *Butyribacterium methylotropicum* [28]. A summary of the mesophillic bacteria that are effective in the fermentation of synthesis gas is given in Table 4.1.

Conversion of synthesis gas to ethanol by *C. ljungdahlii* has been developed into a commercial process that combines cellulose gasification, synthesis gas fermentation, and recovery of ethanol from the reaction products by distillation. The limited solubilities in water at about 37° of carbon monoxide (0.03 g/kg) and hydrogen (0.0014 g/kg) limit conversion rates and a high degree of agitation and increased pressure are necessary to stimulate gas/liquid mass transfer rates. Continuous stirred tank reactors (CSTRs) with high impeller speeds are used to reduce bubble size and so improve interfacial contact and, consequently, reaction rate. A variety of trickle bed reactors have been considered, and these have been more efficient than either CSTRs or bubble column

$$\text{(a) } CH_4 + \tfrac{1}{2}O_2 \ \text{------}\blacktriangleright\ CO + 2H_2$$

$$\text{(b) } CH_4 + H_2O \ \text{------}\blacktriangleright\ CO + 3H_2$$

Figure 4.1 Partial oxidation (a) and steam reforming (b).

$$\text{(a) } CO + H_2O \ \rightleftharpoons\ CO_2 + H_2$$

Figure 4.2 Water gas shift reaction.

Table 4.1 Mesophillic Bacterial Fermentation of Synthesis Gas [36]			
Species	T, °C	pH	Products
Clostridium autoethanogenum	37	5.8−6.0	Acetate, ethanol
Clostridium ljungdahlii	37	6	Acetate, ethanol
Clostridium carboxidivorans	38	6.2	Acetate, ethanol, +
			Butanol, butyrate
Oxobacter pfennigii	36−38	7.3	Acetate, butyrate
Peptostreptococcus productus	37	7	Acetate
Acetobacterium woodii	30	6.8	Acetate
Butyribacterium methylotropicum	37	6	Acetate, ethanol,
			Butyrate, butanol

reactors. Elevated pressures are obviously desirable for bacteria that can withstand pressures of 40−50 Mpa. Although higher fermentation temperatures should lead to improved rates of reaction, they have the downside effect of lowering the solubility of the synthesis gas.

Fermentation of synthesis gas offers a more narrowly defined product slate than does chemical conversion where ethanol, isobutanol, acetates, and butyrates are the primary products. Synthesis gas fermentation also proceeds at lower temperatures and pressures than an analogous FT process, and is apparently independent of a specific ratio of carbon monoxide to hydrogen. For the production of acetates, ethanol, butanol, and butyrates, the synthesis gas fermenting microorganisms depend on the acetyl coenzyme-A pathway that is present in bacteria (Figure 4.3).

Oxidation of H_2 to $2H^+$ or of CO with H_2O to CO_2 and $2H^+$ provides reducing equivalents for conversion of CO_2 to formate, and of

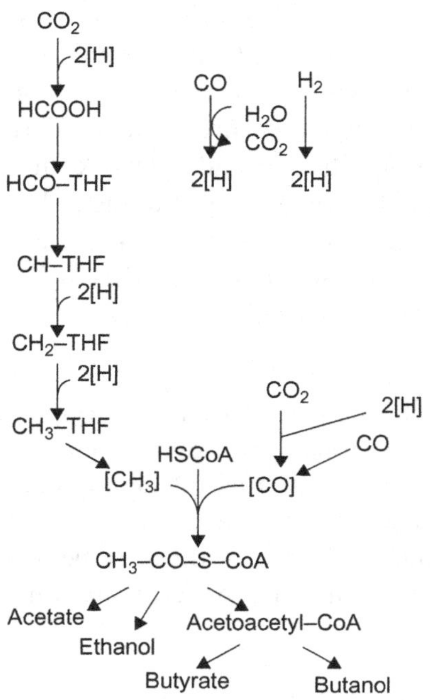

$$CO_2$$
$$\downarrow 2[H]$$

$$HCOOH \qquad CO \qquad H_2$$
$$\downarrow \qquad \begin{array}{c} H_2O \\ CO_2 \end{array} \downarrow$$
$$HCO\text{–}THF \qquad 2[H] \qquad 2[H]$$
$$\downarrow$$
$$CH\text{–}THF$$
$$\downarrow 2[H]$$
$$CH_2\text{–}THF$$
$$\downarrow 2[H] \qquad CO_2 \qquad 2[H]$$
$$CH_3\text{–}THF \qquad \qquad CO$$
$$HSCoA$$
$$[CH_3] \qquad [CO]$$

$$CH_3\text{–}CO\text{–}S\text{–}CoA$$

Acetate   Acetoacetyl–CoA
Ethanol
Butyrate   Butanol

Current opinion in Biotechnology [28]

Figure 4.3 Acetyl-CoA mechanism.

methylene tetrahydrofolate (CH-THF) to methenyltetrahydrofolate (CH$_3$-THF), and of CO$_2$ to CO. Acetyl-CoA synthase/CO dehydrogenase catalyzes the formation of acetyl-CoA from a bound methyl group, a bound CO group, and coenzyme A (CoA).

In bacterial systems, CO$_2$ is first reduced to formate which is then activated with ATP form a formyl group that is bound to the pterin tetrahydrofolate. Acetate is formed from acetyl-CoA to recover metabolic energy. Further reduction of acetate yields ethanol. Butyrate or butanol comes from two acetyl-CoA molecules.

Prominent firms engaged in the bioconversion of synthesis gas include Lanzatech, Ineos, and Coskata.

4.1 LANZATECH

Lanzatech (Auckland, New Zealand) bought at auction the former 100 million gallon Range Fuels plant at Soperton, GA. The plant had never operated despite a loan of $38 million from the USDA and a $43 million grant from the U.S. Dept. of Energy. As the new owner, Lanzatech plans to use the Range Fuels gasifier to produce synthesis gas followed by the use of their own proprietary microbes to produce ethanol and 2,3-butanediol. They intend to work Invista of Wichita, KS, to develop a process to convert the butanediol to butadiene for the chemical market. In addition, they have formed a partnership with Cordon Blue of California whose technology they will use at Soperton. The Cordon Blue system reportedly converts cellulosic wastes to synthesis gas.

In addition, Lanzatech is associated with Virgin Atlantic Airlines for the construction of a demonstration plant for jet fuel in Shanghai, China. Rather than operate a gasifier at Shanghai, Lanzatech will employ CO/hydrogen waste gases from a nearby steel plant. In this manner, the Shanghai plant will not in any way be a biobased facility other than the use of microorganisms for fermentation.

4.2 INEOS

Ineos has had planned an 8 million gallon plant for start-up in 2013. Their technology is based on the gasification of agricultural wastes, such as palm trees. They intend to ferment the synthesis gas to ethanol using *C. ljungdahlii*. Surplus gas will be converted to electricity for a byproduct credit against the process. Ineos is a young British firm ($43 billion in sales) that has grown rapidly through a series of acquisitions.

4.3 COSKATA

Coskata (Warrensville, IL) originally engaged in the gasification of wood chips followed by fermentation of synthesis gas to ethanol and acetate. Its three patent applications focus on novel enzyme systems that are modifications of *Clostridium* bacteria. Although they had a working agreement with General Motors to test the quality of their ethanol, they have decided that synthesis gas can be more economically produced by steam reforming of natural gas and have therefore abandoned near-term development of biomass gasification in favor of natural gas. As a result they have discarded plans to build biomass-to-ethanol plant in favor of a natural gas-based facility. The only relation to biomass in Coskata's efforts are the enzymes used in their fermentation of synthesis gas.

Coskata's decision to employ natural gas in place of waste cellulose is in apparent recognition of the growing supply of natural gas in the United States. Such growth is driven by hydraulic fracturing of shale to the extent that proven reserves have grown dramatically from about 175 trillion cubic feet in 2000 to 317 trillion cubic feet in 2010. The American Gas Association projects that at current rates of consumption of 24 trillion cubic feet per year, the United States has about 100 years of natural gas supply. Commensurate with this growth in supply is a decline in natural gas wholesale price to about $3.50 per million BTUs at Henrt Hub, LA, in early 2014. At this price, natural gas is somewhat cheaper than cellulosic biomass at $3.60–$4.27 per million BTUs. The cellulosic price estimate is based on that of dry wood chips that sold for $60–70 per metric ton for the last 10 years. With increased demand for wood chips, a corresponding increase in their price is likely.

In view of the above and since natural gas can be readily shipped in volume by pipeline, it is difficult to understand the current emphasis on cellulosic biomass as a feedstock for liquid fuels. It would appear that steam reforming of natural gas followed by direct microbial conversion of it to diesel fuel and gasoline would be more economical. Furthermore, why embrace biomass and biosynthesis at all? Once synthesis gas is obtained from natural gas, why not convert it to liquid fuels by established techniques, such as FT technology, catalytic homologation of carbon monoxide, or the ExxonMobil process. A number of companies are already engaged in the development of such efforts. Among them are Enerkem, Sundrop Fuels, Fulcrum Bioenergy, and Syntec Biofuels.

4.4 SYNTEC BIOFUELS

This Vancouver, BC, company uses a fixed bed reactor to produce a combination of methanol, ethanol, n-propanol, and n-butanol by catalytic conversion of synthesis gas using a palladium catalyst in conjunction with oxides of Zn, Ce, and Li. They describe their system as the B2A process. It is described in US patent application 7384987 that was filed on May 25, 2005. They claim the production of 105 gallons of mixed alcohols per ton of cellulose. The firm, which is incorporated in Washington State, had expected to have had its first commercial plant onstream by third quarter, 2011, but there have been no reports that it had done so by early 2014.

4.5 SUNDROP FUELS

This Longmont, CO, firm uses an ultra high temperature heat transfer process in a proprietary RP Reactor for the gasification step (Figure 4.4). The reaction proceeds 20 times faster than conventional techniques. At operating temperatures up to 1300°C, cellulose can be gasified almost instantly to yield a 2:1 ratio of hydrogen to carbon monoxide. Sundrop began construction of a 40 million gallon demonstration plant in 2012 and plans to use the technology in a 3500 barrel per day gasoline plant at Alexandria, VA. They are partners with Chesapeake Energy and plan

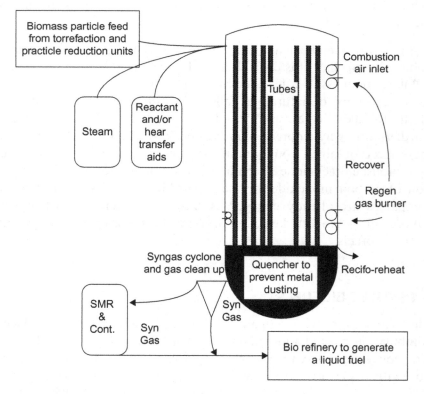

Figure 4.4 Sundrop RP Reactor [30].

to use a variation of the ExxonMobil MTG process in association with Primus Green Energy at Hillsborough, NJ. The proposed NJ plant will produce 93 octane gasoline. Details of the ExxonMobil MTG process are provided later in this chapter. Sundrop's own technology was developed by and licensed from the University of Colorado (Boulder) and the National Renewable Energy Laboratory (NREL).

4.6 ENERKEM

Based in Montreal, this firm plans to build a 10 million gallon ethanol plant in Varennes, Quebec, with the help of a $27 million grant from the Quebec government. The process employs gasification of waste cellulose and includes a first step in which methanol is produced from synthesis gas. The methanol is converted to methyl acetate and acetic acid by rhodium catalysis. The acetic acid is esterified to additional methyl acetate and then hydrogenated to ethanol. Apparently, the methanol need not be produced onsite and can be purchased. Additional plants are planned for Edmonton, Alberta and Pontotoc, MS.

4.7 FULCRUM

Fulcrum Bioenergy of Pleasanton, CA, uses a proprietary process in which synthesis gas is catalytically converted to ethanol. They plan to have their first commercial plant (10 million gal/year) in 2015 outside of Reno, NV. Fulcrum uses a downdraft partial oxidation gasifier followed by plasma arc to produce a carbon dioxide containing synthesis gas. Excess carbon dioxide is removed to obtain the optimum CO/H_2 ratio in what appears to be a chemical catalytic, not enzymatic, system. Propanol and ethanol are produced as is electric power as a byproduct credit. The technology is undisclosed insofar as patents are concerned. Its source is an agreement with Napawin Biomass Ethanol New Generation of Saskatchewan, Canada.

4.8 MAVERICK SYNFUELS

Maverick Synfuels, which was originally named Maverick Biofuels, is located in Research Triangle Park, NC, and uses a combination of synthesis gas-based processes that rely on either steam-methane reforming or gasification of agricultural or municipal wastes. Their system appears to employ a number of existing technologies in which methanol or synthesis gas is converted directly to C_2–C_5 olefins. Their overall technology is referred to as the Olefinity process. A pair of U.S. patents (8,354,563 and 8,344,188) describe the conversion of methanol to olefins by contacting them with a silicoaluminophosphate catalyst at 250°C under 1000 psi pressure. The olefins can be subsequently oligomerized either directly or through their corresponding alcohols to diesel fuel. The work is similar to efforts in the 1970s by Union Carbide Corp. (now Dow Chemical) and Mobil Oil (now ExxonMobil).

4.9 EXXONMOBIL MTG PROCESS

The ExxonMobil MTG process was introduced in 1975 as a means to convert methanol to gasoline. The initially published US patent (US 4,404,414) in 1983 has been followed by at least 30 more, largely assigned to Mobil Oil prior to its merger with Exxon Corp. The methanol is obtained by steam reforming natural gas to synthesis gas. In the process, methanol is first converted to dimethyl ether by being passed over a fixed bed gamma-alumina at pressures of 15–175 psig. The conversion is very exothermic with 750 BTUs of heat produced per pound of methanol. The methyl ether is then passed over a fixed bed shape selective ZSM-5 zeolite to produce liquid fuels in 75% selectivity at 100% conversion. Accordingly, 90% of the carbon in methanol is converted to gasoline. The liquid fuels are C_{12} and lower which is optimal for the gasoline range. The 25% gaseous products are largely C_3s and C_4s and are in the proper balance for higher octane gasoline. Pilot plant data indicate that 1.5 barrels per day of gasoline were obtainable per barrel of methanol. A commercial plant based on low-cost natural gas was operated in New Zealand from 1986 to 1996 and was finally shut down when New Zealand area natural gas supply was no longer competitive with world crude oil prices. A variation of the ExxonMobil MTG process is contemplated by Primus Green Energy who have constructed a 100,000 gallon demonstration plant at Hillsborough, NJ. The synthesis gas to be used is a combination of municipal solid waste and natural gas. This is discussed later in this chapter. Similarly, the proposed MTG plant of Sundrop Fuels will be based on gasification of forest wastes supplemented with hydrogen from natural gas. The interest in the chemical catalytic conversion of synthesis gas to gasoline suggests that the fuels industry may have come full circle from the time that synthesis gas from either natural gas or coal gasification were economic alternatives to crude oil. In this instance, it is cellulosic waste that is the alternative raw materials.

4.10 FISCHER—TROPSCH

The resurgence of interest in the MTG process, either biomass or fossil fuel based, is further illustrated by a 100,000 ton plant that is under construction by the Jincheng Anthracite Mining Group in China. This party uses coal gasification. The first stage of the plant started up in 2009 and a second stage of 1 million tons capacity is planned. The first license to a coal-based facility was granted to DKRW Advanced Fuels of Medicine Bow, Wyoming.

The fermentation of synthesis gas to lower alcohols is formally similar to the well-established FT process that catalytically achieves similar results. In the 1920s, Fischer and Tropsch announced a process in which an iron catalyst effected the condensation of carbon monoxide and hydrogen at atmospheric pressure over an iron catalyst to produce a mixture of alcohols, aldehydes, ketones, and fatty acids. Improvements in catalyst type which varied with reaction pressure and the use of fixed or fluid bed catalyst systems led to products in the gasoline range and higher. The results from such improvements in comparison to the ExxonMobil MTG process are shown in Table 4.2. FT technology was used in Germany during World War II and later in South Africa in the 1970s and later when access by either country to crude oil was limited politically. In each instance, local coal reserves were abundant and were turned to for synthesis gas production since crude oil was unavailable to either Nazi Germany or the Union of South Africa. In the latter case, it was because of disapproval of Apartheid that was practiced in that

Table 4.2 Percent of Products: FT Versus ExxonMobil MTG			
	FT Fixed Bed	FT Fluid Bed	MTG
C_1-C_2 light gases	7.6	20.0	1.3
LPG (C_3-C_4)	10.0	23.0	17.8
Gasoline (C_5-C_{12})	22.5	39.0	80.9[a]
Diesel ($C_{12}-C_{18}$)	15.0	5.0	0
Heavy oil ($C_{19}+$)	41.0	6.0	0
Oxygenates	3.9	7.0	0
[a]G. A. Mills, Chem. Technol. 7, 418 (1977); Kirk Othmer Encyclopedia of Chemical Technology, 11, 477.			

country. Coal-based FT plants are still operated in the Union of South Africa.

If the intent of a biomass producer is to make automotive fuels, the MTG process is far more attractive since the production of diesel and gasoline together total 88.7% compared to 37.5% for FT technology. Supposedly, the fixed investment cost for the MTG plant would be considerably less since for the same volume of automotive fuels, the FT plant must be considerably larger. The broad product distribution from a FT plant is illustrated in Figure 4.5. Efforts to enhance gasoline yield by oligomerizing the reactants to greater chain lengths ultimately result in yield loss because of wax production.

A Royal Dutch Shell FT project in Qatar has a fixed investment construction cost of about $20 billion for a 140,000 barrel per day plant. The plant produces aviation jet fuel for Qatar Airways. The fuel is a blend of 50% synthetic paraffinic kerosene produced by the plant. For those firms considering an analogous effort based on waste cellulose, considerable credit for its nonautomotive fuel byproducts must be sought in order to offset its probably large capital investment. The ExxonMobil MTG process is more narrowly focused on the production of automotive fuels. The question then becomes: which is less costly, the investment and operating cost for the *enzymic* conversion of synthesis gas to fuels versus the catalytic version of the ExxonMobil MTG process? The jury is still out with firms presently pursuing both avenues of commercialization. There is an advantage to the MTG system in terms of operating experience.

An option still remains for the application of FT process to the gasification of cellulose. The first and only commercial FT plant in the United States was started up in 1950 by Carthage Hydrocol in Brownsville, TX. It was designed to convert 90 million standard cubic

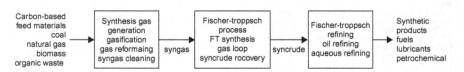

Figure 4.5 Generic Fischer Tropsch Process. Copyright reserved: Sasol Technology R&D, FTR and C1 Chemistry Research, HCC 19 August 2010.

feet of gas per day of natural gas that was cost competitive at the time. The products were to be 6000 barrels per day of gasoline, 1100 barrels per day of diesel, and fuel oils plus 300,000 lbs of chemicals. Poor yields, mechanical failures, and equipment problems plagued the installation such that it never attained more than 30% of nameplate capacity. In 1953, the facility was taken over by Amoco, and it was shut down permanently in 1957. Interest in FT nevertheless continues actively with a 12,000 barrel per day plant for diesel being operated by Shell Oil in Bintublu, Malaysia, and a quite new 140,000 barrel per day unit also by them in Ras Laffan, Qatar. Both facilities employ regionally available inexpensive natural gas. In late 2013, Shell also contemplated the construction of another FT plant in Louisiana, but canceled the project before year end.

Conventional FT facilities may be too large to be applicable to waste cellulose. Sundrop and Primus Green Energy, however, and perhaps, Coskata, believe there is an opportunity for catalytic gasification of waste cellulose to fuels. Although the original size of the natural gas-based New Zealand plant for the ExxonMobil MTG process was 100,000 tons per year, these firms believe that a plant of that size is feasible for waste cellulose. Major issues for a cellulose gasification plant using FT technology are its large capital investment and the ability to gather the huge amount of waste cellulose required for feedstock. Gathering natural gas in volume is a far simpler exercise especially when pipelines are available.

Primus Green Energy plans to produce gasoline by the ExxonMobil process from wood biomass by first gasifying it and then converting the synthesis gas to methanol and finally gasoline. Ultimately, they intend to employ natural gas in place of wood waste as the source of synthesis gas. As noted earlier in this chapter, they commissioned a 100,000 gallon biomass-based demonstration plant in October 2013 and they are working toward a 27.8 million gallon commercial plant based on natural gas. Ground breaking for the commercial unit is planned for 2014.

The Primus Green Energy process employs a four-stage reactor system [74] with a single recycle loop that provides the requisite thermal capacity to moderate the high heat release of the reactants and to provide the reactants and environment for the efficient operation of the overall process (Figure 4.6).

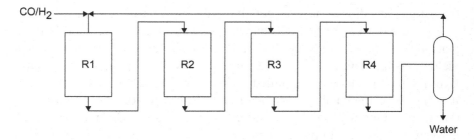

Figure 4.6 Primus Green Energy synthesis gas to gasoline process. Green Car Congress; April 30, 2014.

The first stage (R1) converts part of the synthesis gas to methanol. The second stage (R2) converts the methanol to dimethyl ether to gasoline, the third stage (R3) converts the methanol and methyl ether to fuel. The fourth stage (R4) converts the high melting component (durene) and other low volatility aromatic compounds, such as tri- and tetramethylbenzenes to branched high octane paraffins. The reactions produce water as a byproduct that is carried through the system to a high-pressure separator that is positioned after the fourth stage. The streams from the separator are a liquid fuel stream, a water stream, and a gaseous stream that contains light hydrocarbon reactants and unchanged synthesis gas. The larger part of the gas stream is recycled to the inlet of the first stage and combined with fresh synthesis gas. Alternatively, the fresh synthesis gas stream is mixed with the product from the second stage. The smaller part of the gas stream from the separator is sent to hydrocarbon recovery and to fuel gas to preheat the various streams. The liquid fuel is sent to blending into gasoline, jet fuel, or diesel, and the water goes to the synthesis gas plant for steam generation.

Primus reportedly claims that its system constitutes a breakthrough for the conversion of gas to liquids since it offers attractive economics at a scale <6000 barrels per day. Supposedly, one million BTUs of natural gas can be converted to five gallons of 90 + octane drop-in gasoline. More than 35% by weight of the synthesis gas is converted into final product. Primus further claims the gasoline produced from its proposed commercial plant will be competitive with gasoline produced from crude oil at $65−70 per barrel. Brent crude oil in the first quarter of 2014 was priced at about $100 per barrel.

An interesting FT development that could affect its competiveness is the technology offered by Velocys which was formerly the Oxford

Catalyst Group. This firm has microchannel technology that is able to enhance the FT technology to the extent that a 7.6 million gallon plant may be economical. Such a plant would only require 500 metric tons per day of cellulose. The microchannel technology enables the use of cobalt catalysts without the presence of precious metal promoters without any loss in performance. Microchannel reactors are designed for economic production on a small scale such as would be used in the gasification of cellulose. These reactors are compact and have tube sizes in the millimeter range as opposed to those in the centimeter range that is typical of conventional FT reactors. When the microchannel reactors are used in conjunction with the cobalt catalysts developed by Oxford Catalysts, conversion efficiencies of 70% per pass are possible. A single microchannel reactor block measures $2 \times 2 \times 2$ ft. in size and can produce 30 barrels of liquid fuel per day. In contrast, conventional FT facilities typically have conversion efficiencies of only 50% per pass and are designed to operate at a minimum of 5000 barrels per day. Accordingly, a FT facility that employs microchannel reactors could be attractive for the conversion of waste cellulose into liquid fuels.

A proposed 7.6 million gallon Velocys plant is not much different in size from the new KIOR plant in Columbus, MS. Velocys has a partnership with Ventech Engineers who specialize in the design and construction of modular refineries. Under their agreement, Ventech will design and sell FT plants that incorporate the microchannel reactors of Velocys in North America. An order has been placed for the first module for the first 1400 barrel per day FT plant at Pinto Energy in Ohio. The plant is ostensibly natural gas based. Velocys was formed in 2001 as a spin-off from Battelle Institute, and Oxford Catalysts was originally formed in 2004, as a product of the Oxford Wilson Catalyst Group. The microchannel technology was developed at Battelle Institute. The two companies merged in 2008.

An issue with gasification of waste cellulose is the prospect of pollution caused by some of its components. This has been addressed by Sierra Energy of Sacremento, CA, who have introduced the FastOx process [31]. The system operates on the principle of a modified blast furnace and converts the cellulose to a synthesis gas that is suitable for conversion to liquid fuels or power generation.

The synthesis gas fed to a FT process is essentially independent of its origin. It doesn't matter whether it was produced from coal, natural gas, or waste cellulose. At the trace level, there may be the differences in the contaminants, and this may be affected only by the technique used to clean gas. There might also be differences in the hydrogen/CO ratio but this is dependent on the design of the gasifier and is independent of the starting raw material. The point is that the FT can be used in conjunction with any source of reasonably clean synthesis gas. Production of the latter in a FT plant using fossil fuels is the most costly aspect of the operation and requires a substantial investment in utility infrastructure. The fixed capital investment of synthesis gas generation (partial oxidation or steam reforming) is generally between 50% and 67% of the total capital cost for the plant. Should the cost for gasification of cellulose be significantly less, it would appear that there is an opportunity to produce liquid fuels economically by this means. The configuration of a generic FT plant that might use either waste cellulose or natural gas as a feed, and possibly both in the same plant, is illustrated in Figure 4.7.

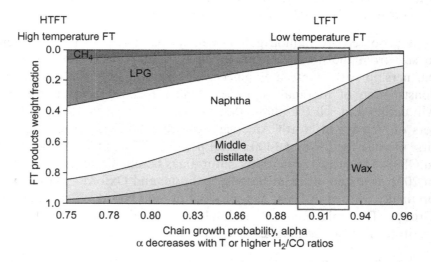

Figure 4.7 Generic FT process.

CHAPTER 5

Isobutanol

Ethanol has long dominated renewable fuels and accounts for 83% of the total. This market dominance appears to be in transition with a decided trend toward isobutanol and diesel regardless of the technology that is used for its production. At the very least, continued growth in the production of ethanol by fermentation of cornstarch may be at an end and may even be declining. As noted in Chapters 3 and 4, most cellulosic processes, especially those that employ gasification, produce butanol as a principal product.

Isobutanol is worth more to refiners than methanol because it has 26% more energy. Unlike ethanol that is 100% miscible with water, isobutanol has limited (8.5%) solubility, and among other advantages, it doesn't cause stress cracking in pipelines. Furthermore, the lower oxygen content of the isobutanol molecule enables greater amounts of it than ethanol to be blended into gasoline. Although stress corrosion cracking of carbon steel is of concern in blended gasoline that contains more than 10% ethanol, it is not an issue with isobutanol contents of 12.5–50%. Consequently, isobutanol can be blended into gasoline as a drop-in fuel at refineries and can readily be shipped by existing pipelines to fuel terminals. The blended octane number of isobutanol of 100 is midway between that of ethanol (112) and gasoline isobutyl alkylate. A comparison [40] of the properties of isobutanol relative to ethanol is illustrated in Table 5.1.

Isobutanol overcomes the blend wall limitations of ethanol by producing a similar grade of gasoline at only a 2.7% oxygen content for refiners. This generates 16.25 RINs per gallon of finished product. For the same volume percent, there is an EPA waiver that would allow up to 16.1% isobutanol and furnish 20.93 RINs per gallon of blended gasoline or more than twice that of E10 for an equivalent oxygen content.

Isobutanol qualifies as an Advanced Biofuel under the RFS2 discussed in Chapter 1. To account for the relative amount of renewable energy benefit that each biofuel generates based on its energy

Table 5.1 Properties of Blended Gasolines [40]

	Volume Pct.	Oxygen Content, %	RIN, gal/gal of Product
E10	10	3.5	10
E15	15	5.2	15
Isobutanol	12.5	2.7	16.25
i-butanol*	16.1	3.5	20.93
*EPA waivers			

Table 5.2 Energy Comparisons of Gasoline [40]

	Ethanol	Isobutanol	Alkylate
Blended octane number	112	102	95
RVP, psi	18–22	4–5	4–5
Oxygen content, %	34.7	21.6	0
BTUs/gallon	84,000	110,000	116,000
Net energy, % of gasoline	65	82	95

content, Advanced Biofuels must have at least a 50% reduction in greenhouse gas footprints. Since isobutanol has a higher energy content than ethanol it therefore provides 1.3 RINs per gallon for each 1.0 gallon for ethanol. The energy advantages of isobutanol are further illustrated in Table 5.2. The much lower Reid vapor pressure (RVP) means that gasoline blended with isobutanol is less volatile than fuel blended with ethanol. As a result, refiners do not have to remove some of the lighter components from their gasoline in order to meet air quality standards. This is even more attractive in such hot weather states as Arizona where volatility is a major issue.

In December 2013, Underwriters Laboratory (UL) further paved the way for the use of isobutanol in US gasoline. UL approved up to 16% for its use in UL 87A pumps by any manufacturer that met ASTM specifications. This approval assures all US service stations that isobutanol blended gasoline in their gasoline current pumps without the need to purchase new equipment.

An early practical demonstration of the utility of isobutanol took place during the summer of 2005 when a 1992 stock Buick Park

Avenue was driven for 10,000 miles around the United States using 100% isobutanol. Not only were there no problems, but the vehicle averaged from 24 to 28 miles per gallon. The same car had an average mileage of 22 miles per gallon on regular gasoline. A further demonstration of isobutanol's practicality took place during the 2012 Olympic games in London when BP fueled several BMW automobiles with isobutanol.

Isobutanol is certainly not a new molecule and has been produced in the petrochemical industry for many years, notably by the OXO process in which propylene is reacted with synthesis gas by cobalt catalysis. Its primary use has been as a solvent and as an intermediate for conversion to isobutylene, which in turn is an intermediate for the production of butyl rubber. The cost of isobutanol production by chemical synthesis has been too great for it to be considered for use as a fuel. What has changed all this is a look back more than 150 years to the work of Louis Pasteur in 1861 that was followed by the efforts of Chaim Weizmann [41] in 1918. He isolated *Clostridium acetobutylicum* that was commercialized to produce acetone/butanol/ ethanol anaerobically in a ratio of 3:6:1 from starch. The process is commonly referred to as the ABE process after the first initials of the products. The acetone was used to produce Cordite, an explosive. Subsequently, Commercial Solvents Corp. operated a number of ABE plants in the United States and United Kingdom from 1920 to 1946. Cost issues, relatively low yields, and slow rates of fermentation in addition to problems caused by end product inhibition, led to the eventual shutdown of the plants as uncompetitive against the petro-chemical synthesis.

The metabolic pathway for the original ABE process is schematically illustrated in Figure 5.1. Glucose (derived from starch) is converted to acetyl Co-A that can optionally be converted to acetate and ethanol. The acetate can be reconverted to additional Co-A or react with it to form acetoacetyl Co-A. The acetoacetyl Co-A can then go on to form acetoacetate which decomposes to acetone. The acetoacetyl Co-A goes on to form butyryl Co-A and finally isobutanol.

Ketol-acid reductoisomerase (KARI) enzymes are ubiquitous and are involved in the production of valine and isoleucine and also affect the biosynthesis of isobutanol. The latter is specifically produced by

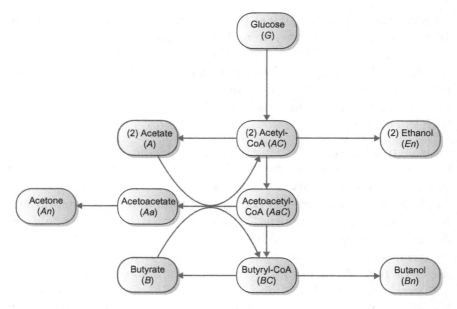

Figure 5.1 Acetone−butanol−ethanol fermentation [62].

the metabolic breakdown of L-valine in yeast fermentation. Isobutanol is a component of the "fusel oil" that results in the incomplete fermentation of amino acids by yeasts. After the amine group of L-valine is harvested as a nitrogen source, the resulting alpha-ketoacid is decarboxylated and reduced to isobutanol. Addition of exogenous L-valine and other amino acids serves to increase the yield of isobutanol and C_3 and C_5 alcohols when they are added to the fermentation medium. Since the use of valine and other amino acids on an industrial scale is prohibitively expensive in the production of isobutanol, Butamax has developed a more cost-effective technique. They have developed a method for producing isobutanol directly from pyruvic acid, a common product in the metabolic breakdown of glucose.

Microorganisms such as yeast and bacteria are capable of producing isobutanol through a five-step sequence such as the following: (i) pyruvate to acetolactate, (ii) acetolactate to 2,3-dihydroxyisovalerate, (iii) 2,3-dihydroxyisovalerate to alpha-ketoisovalerate, (iv) alpha-ketoisovalerate to isobutyraldehyde, and finally (v) isobutyraldehyde to isobutanol. These are shown schematically in Figure 5.2.

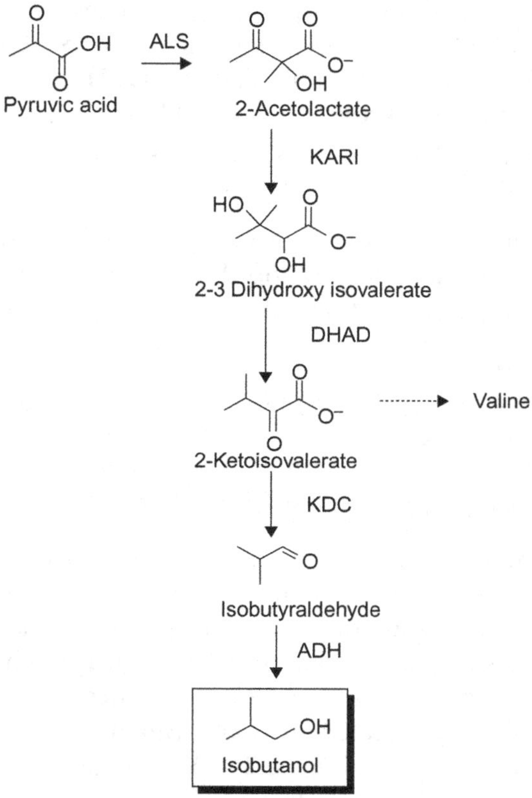

Figure 5.2 Isobutanol metabolic pathway. The improved method of Butamax consists of the introduction of engineered deoxyribonucleic acid (DNA) into microorganisms in order to stimulate isobutanol production.

The engineered DNA constructs encode enzymes that catalyze each of the five coupled reactions and consequently the overall rate and yield of isobutanol formation.

Butamax argues that the only KARI enzymes that were known are those that bind NADPH in their native form, and that these reduce the efficiency of isobutanol production. They argue that a KARI enzyme that would bind NADH would be beneficial and therefore enhance isobutanol synthesis by capitalizing on the NADH produced by the existing glycolytic and other metabolic pathways most commonly used in microbial cells. They further argue that the discovery of a KARI enzyme that can use NADH (reduced nicotinamide adenine dinucleotide) as a cofactor would be a significant advance in itself. This position is at the heart of their infringement litigation with

Gevo that is discussed later in this chapter. To achieve this, the development of a KARI with a great specificity for NADH as opposed to NADPH is essential. NADH and NADPH are largely similar molecules except for the presence of an additional phosphate group in the 2-position of the ribose bonded to the adenine segment of the NADPH molecule.

Since isobutanol as it forms can be toxic to the KARI enzyme system, Butamax [43] has also developed a technique for its continuous removal by liquid—liquid extraction in the course of fermentation.

Some of Clostridium's limitations can be overcome by the introduction of its isobutanol pathway into organisms that grow faster, that can tolerate high concentrations of isobutanol, or can metabolize alternate feedstocks. *E. coli*, for example, is one example of such a microorganism that has a high isobutanol rate of growth [44].

Butamax and Gevo are the two most prominent organizations that are developing superior fermentation systems for isobutanol. As cited earlier, their efforts focus on the optimization of the now classical ABE enzyme system discovered by Chaim Weizmann in Britain in 1916. Of the two industry leaders, Gevo, located in Douglas, CO, is the smaller and was founded in 2005. Butamax is a joint venture between BP and DuPont and located in Wilmington, DE. Their objective as a company is the manufacture and licensing of isobutanol as an alternative motor fuel. Both Gevo and Butamax are cognizant of the opportunity to produce isobutanol by retrofitting cornstarch-based ethanol plants. Each has its own extensive patent portfolio, and to a large degree, overlapping technologies based on engineered KARI enzymes. As would be expected, they have been locked in patent infringement litigation for at least 2 years. The litigation has embraced suits and countersuits of each others patent portfolios. The most recent court decision was in July 2013 and held in favor of Butamax.

In May 2012, Gevo began to produce isobutanol at their retrofitted 18 million gallon ethanol plant in Luverne, MN. The plant quickly encountered operating problems and was subsequently shut down to address them. In the interim, the plant returned to the production of ethanol. The plant was restarted for isobutanol production in May 2013. As in the case of Butamax, an isobutanol product stream is continuously taken from the fermentation medium to maintain concentration below the level of toxicity to the organisms. It is subsequently returned for isobutanol production. This practice allows for improved conversion to isobutanol. By retrofitting an ethanol plant such as this, the project time for starting up a new isobutanol facility is greatly reduced compared to a grassroots plant. Accordingly, the fixed capital investment for a retrofitted plant can be as much as 20−40% lower than for a new on-purpose one. Reportedly [40], operating costs and utilities are otherwise comparable to cornstarch-based ethanol production. This suggests that the cost for isobutanol production should otherwise be similar to that for ethanol.

Gevo has also developed a process for employing isobutanol produced in its plants for jet fuel production [45]. This encompasses a successive series of steps as portrayed in the plant configuration in Figure 5.3 and is summarized as follows:

- Dehydration of isobutanol to isobutylene
- Oligomerizarion of isobutylene to $C_{12}-C_{16}$ trimers and tetramers
- Hydrogenization to jet fuel.

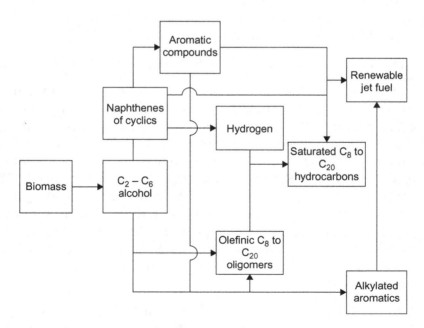

Figure 5.3 Gevo jet fuel process configuration. US Patent 8,378,160.

The reactions are well established and have been used commercially by others. The jet fuel system has been run in a 120,000 gallon demonstration unit, and the product has been sold to the US Air Force as part of the Alternative Fuels Certification process.

In April 2014, Gevo alerted its investors that it might not be able to obtain the financing it needed to continue operations through the end of 2014. The company incurred a loss of $67 million in 2013 and had accumulated a deficit >$262 million. This raises doubts on their ability to continue as a viable concern. Apparently, the struggle in 2012 to convert its Luverne, MN, plant from ethanol to isobutanol and its lengthy legal battle with Butamax over patent infringement have exacted a heavy toll.

In June 2006, DuPont and BP formed a partnership for lignocellulosic feedstock conversion to isobutanol. Three years later, the partnership acquired Biobutanol LLC, a US firm, and subsequently formed Butamax Advanced Biofuels, as mentioned previously, for the licensing of their isobutanol technology. They also formed a Butamax Early Adopters Group of which Highwater Ethanol was the first of seven members in early 2013. Butamax also entered into a collaboration agreement in 2012 with Fagen Engineering of Granite Falls, MN, to engineer the retrofitting of ethanol plants to isobutanol.

Construction was initiated in 2014 for a Highwater Ethanol plant.

In late 2012, DuPont itself began the construction of a 30 million gallon cellulose-to-ethanol plant in Nevada, IA, for start-up in 2014. The 210 million gallon plant will be based on corn stover and will use data from a demonstration plant that DuPont has operated in Vonore, TN, since 2008. The technology to be employed is ostensibly the engineered KARI enzymes that Butamax developed. BP, DuPont's partner in Butamax, plans to construct a 20,000 gallon isobutanol demonstration plant at Hull, UK, in conjunction with their 100 million gallon ethanol plant at the same location. The proposed ethanol unit will use sugar cane, wheat, or cornstarch as raw materials. BP and DuPont plan to convert this facility to isobutanol eventually.

Prior to the advent of opportunities as a fuel, producers sought to limit isobutanol production by yeasts since its presence can ruin the flavor of wine and beer. Typically, yeasts produce isobutanol in two separate locations within the cell. As noted earlier, the synthesis begins with pyruvate that is formed in metabolic breakdown of glucose. The pyruvate is transported into the mitochondria where it can enter many different metabolic pathways, including the one that results in valine production. Alpha-ketoisovalerate, a precursor for both valine and isobutanol, is produced in the mitochondria. Valine and alpha-ketoisovalerate can then be transported out to the cytoplasm where they are eventually converted to isobutanol. Efforts at MIT [46] found that if the final steps were moved from the cytoplasm into the mitochondria, the researchers were able to enhance isobutanol production by 260% and the yields of two related alcohols, isopentanol and 2-methylbutanol, by as much as 370% and 500%, respectively. Long term, this suggests the possibility that these two alcohols may be candidates for renewable fuels.

The fermentation of synthesis gas to ethanol as the primary product is undergoing active development (see Chapter 4). Now, Syngas Biofuels Energy of Houston, TX, is actively engaged in the use of such technology for isobutanol production [47]. In their approach, the genes in *Clostridium* species MT871 are engineered for isobutanol biosynthesis. Furthermore, these enzymes are tailored to withstand 6.1% isobutanol in the fermentation broth in order to prevent product inhibition during continuous fermentation of synthesis gas. The firm speculates an isobutanol cost of production 53.94 cents/gallon including a 100% return on capital after 1 year of operation. This is somewhat greater than the 2014 market price of 45 cents/gallon for isobutanol. The total investment cost for the hypothetical 30 million gallon plant is estimated to be $56 million.

A number of firms are engaged in *n-butanol* production by the use of engineered microorganisms. Because of its lower octane numbers, straight chain n-butanol is less attractive than isobutanol as a blending agent for gasoline. A comparison of the properties of n- and isobutanol is given in Table 5.3.

The average of the two octane numbers, $(RON + MON)/2$, is the target commonly employed to determine the attractiveness of a fuel. Linear butanol is clearly less attractive for motor fuel. As a result, most efforts in n-butanol production from engineered microorganisms are focused on its eventual use in chemical production. An alternate

Table 5.3 Properties of n- and Isobutanol		
	n-Butanol	Isobutanol
RON	94	113
MON	78	94
(RON + MON)/2	86	104
Boiling point, °C	108	117
Vapor pressure @37.8°C, kPa	3.9	2.4
Solubility in water, wt.%	7.7	8.5

technique under development is the Guerbet reaction which comprises nobel metal-catalyzed coupling of ethanol. A fully developed process for n-butanol is rhodium-catalyzed carbonylation of propylene.

Chemical applications for n-butanol include uses in solvents and the production of PVC plasticizers such as dibutylphthalate.

CHAPTER 6

Algae as a Fuel Source

Algae is ubiquitous and comprises two main species, micro- and macro- of which seaweed is a typical example. Macroalgae are true plants and absorb nutrients through their cell structure, and because of their size, can be readily harvested. Microalgae, to the contrary, are a microscopic species and are favored universally for the development of renewable fuels because they grow far more rapidly than macroalgae and because they have a much greater lipid content. Microalgae grows virtually everywhere in fresh or in salt water. Every owner of a swimming pool has had to combat the presence of microalgae at one time or other. Microalgae, as their name implies, are unicellular organisms that exist individually or clustered into chains. Their sizes range from a few to several hundred micrometers. Unlike higher plants, microalgae lack leaves or roots although they are capable of engaging in photosynthesis. Because of this, they produce about half of the earth's atmospheric oxygen and simultaneously they consume half of its carbon dioxide in order to enable their own growth. This double-barreled photosynthetic action makes them desirable for the reduction of greenhouse gases while providing a means for the production of renewable fuels. In the following discussion, the term, algae, will be used with reference to microalgae.

It has been estimated that about a half million species of algae exist of which about 50,000 have been described. Of these, more than 15,000 novel compounds have been chemically determined. Algae are composed of varying amounts of protein, carbohydrates, fats, and nucleic acids. Although the percentages of these vary by species, many types are composed of up to 48% by dry weight of fatty acids. It is this fatty acid (or oil) content that can be extracted for conversion to biodiesel that had attracted the attention of considerable research. The fatty acids are linear, branched, saturated, or unsaturated. A short list of the variation in lipid content of some algae species is illustrated in Table 6.1.

Table 6.1 Oil Content of Algae Species	
Species	Oil Content, % Dry Weight
Batryococcus braunii	29–75
Chlorella	29
Chlorella protothecoides	15–55
Cyclotella DI-35	42
Hantzschia DI-160	66
Crypthecodinium cohnii	20
Neochloris oleobundans	35–54

A basic question is which species is best for biodiesel production? The answer to this was pursued by the NREL, a branch of the U.S. Department of Energy, that is focused on renewable energy and efficiency. NREL operated an Aquatic Species Program that was directed to the identification of lipids from algae. Their effort was completed in 1998 and concluded that the optimum algae for fuels are Chlorophyceae, so-called green algae, that have very high growth rates at 30°C. *C. muelleri*, for example, doubles in growth rate over a temperature range of 20–35°C and can exhibit as much as a fourfold daily increase.

The other species favored by NREL are diatoms. These differ from green algae in that they require silicon in water to grow whereas green algae require nitrogen. Interestingly, algae are known to accumulate more lipids under nutrient deficient conditions. Diatoms are unicellular organisms and are the most common type of phytoplankton. Such algae have very large amounts of C_{20} polyunsaturated fatty acids, such as eicosapentaenoic acid, that are useful for diesel synthesis. The major fatty acids in Chlorella are saturated and unsaturated C_{17} compounds. Taken as a group, algae have an extraordinarily large amount of natural oils compared to other common sources of oils for biofuels production as shown in Table 6.2.

The production of diesel and other liquid transportation fuels from lipids in algae has been under investigation on bench and pilot plant scale for many years going back to 1953. The initial effort was directed toward protein production for food. Many issues, such as culture

Table 6.2 Lipid Production by Source	
Source	Gallons of Oil/Acre/Year
Corn	18
Soybeans	48
Safflower	83
Sunflower	102
Rapeseed	127
Palm	635
Algae	5000–15,000

contamination and the harvest of algae by centrifugation or settling, deterred development. Early work by Germany during World War II deterred development for the same reasons. Denied access to ample crude oil during the war moved them in the direction of coal based FT synthesis as discussed in Chapter 4. The Germans discovered enhanced production of Chlorella under reduced nitrogen pressure as did NREL subsequently. Chlorella laboratory yields of 220 g (dry) per cubic meter in 14 days were obtainable. This could be equated to 50 g/square meter/day in open ponds. In 1953, Arthur D. Little Inc., the former consulting company, reported pilot-scale production of Chlorella of 11 g/square meter/day over 10-day periods in 300 square feet of glass tubes. About the same time, workers at the Carnegie Institution reported similar results in four shallow, nonagitated open ponds that were lined with a polymer coating.

Recently, an economic analysis [48] at Kansas State University of diesel production of algae grown in open ponds concluded that diesel fuel obtained in this manner was impractical. A volume of 50 million gallons would require about 11 square miles of open ponds. Such acreage for a plant of this size would supply only 0.1% of the US market. The hypothetical plant was based on transesterification of an algal oil with methanol to produce a methyl alginate of 292 g molecular weight. The fatty acid composition was modeled after that of soybean oil and contained 6%, 52%, 25%, 5%, and 12%, respectively, of linolenic, linoleic, oleic, stearic, and palmitic acids.

Concerns about the sustainability of open ponds for algae to fuels production were also echoed by the National Research Council. In their report [49], the production of sufficient algal-based fuel to meet at least 5% of US transportation fuel demand (ca. 39 billion liters) would place unsustainable demands on energy, water, and nutrients. These conclusions suggest the preferable use of closed reactor systems as opposed to open ponds for the commercial use of algae for fuel. Such closed reactor systems are now being used by a number of parties and are commonly referred to as photobioreactors (Figure 6.2). In these closed systems, algae and all its growth requirements are introduced under controlled conditions as for any chemical or commercial fuel process. PBRs may require 10 times the capital investment for an open pond system [103]. Reportedly, the estimated production cost for algae in an open pond system is $10/kg compared to $30−70/kg for a PBR. If one assumes that algae contains 30% oil by weight and there is no cost for carbon dioxide, the cost for algae oil produced from a pond and a PBR would be $1.81 and a $1.40/l, respectively. The corresponding cost for diesel produced from each of these systems is dependent upon the process that would be used for each. A general description of a photobioreactor [50] process system is provided in Figure 6.1.

With reference to Figure 6.1, algae goes from a feed tank to a diaphragm pump that moderates the flow into a polyacrylate tubular reactor. The pump contains a carbon dioxide inlet valve. The reactor itself is used to promote algae growth by controlling the introduction of process variables including light. The reactor also has a built-in purge system that cleanses it without stopping the process. Once the

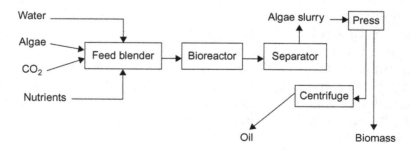

Figure 6.1 Bioreactor process system.

algae has completed its passage through the reactor, oxygen sensors monitor oxygen build-up and enable the release of oxygen from the storage tank. At this stage, another sensor measures the optical density of the cells in order to determine their rate of formation. The algae is finally passed through a filtration system for recovery. Any unrecovered algae is recycled to storage.

A number of advantages accrue to the use of a PBR follows:

1. The algae can be grown under controlled conditions allowing for greater productivity
2. The system has a large ratio of surface-to-volume and thereby greater efficiency. PBRs provide maximum efficiency in the use of light. Typically, the algae cell density is 10−20 times greater than in alternate systems
3. Better control of gas transfer
4. Reduction in evaporative loss
5. Uniform process temperatures
6. Good protection against outside contamination
7. Space savings. The PBRs can be installed vertically, horizontally, or at an angle
8. Tube self-cleaning mechanisms can dramatically reduce fouling of the PBR.

In addition to the use of PBRs, some parties, such as Iowa State University [51] have engaged in enzyme engineering to further increase algae productivity. They report that algae biomass can be increased 50−80% by the combination of two engineered genes that engage in photosynthesis. The genes are described as LC1A and LC1B. The combined genes perform well in carbon dioxide-rich environments in which genes would otherwise shut down. The work was done with *Chlamydomonas reinhardii*. When used separately LC1A and LC1B offered a notable but far less enhancement of only 10−15% in biomass increase. In achieving the productivity enhancement, the Iowa State group was also able to produce fatty oils instead of starch that is the usual result in productivity increases.

Unitel Technologies of Mt. Prospect, IL, describes a process for lipid recovery from algae that bypasses two energy intensive steps: (i) removal of water from algae and (ii) extraction of the lipids for recovery. They employ a specialized hydrolysis reactor from which the fatty acids can be recovered directly. The algae is sent to the reactor that is maintained under 1050 psig hydrogen pressure and at 270°C. Under these conditions, the fatty acids can be readily recovered by phase separation. The fatty acids remain in an upper phase and the byproduced glycerol is found in the lower (aqueous) phase. Recycled fatty acids act as catalysts to facilitate the hydrolysis.

In a press release on December 17, 2013, Pacific Northwest National Laboratory (PNNL) discloses an end run to produce fuel from algae. In contrast to efforts by others to recover either fatty acids or triglycerides as intermediates for fuel, they report a technique for converting algae to crude oil within minutes. The system operates under near to critical conditions of 350°C and 3000 psi. The reaction conditions combine hydrothermal liquefaction and catalytic hydrothermal gasification. Conversions of algae to crude oil of >90% occur in <1 h. Byproducts are a phosphorus containing material that can be recycled to grow more algae. Supposedly, the system is designed to produce algae in open ponds as well as in PBRs.

PNNL is an arm of DOE and is managed by Battelle Memorial Institute. They have been collaborating with Genifuels of Salt Lake City, UT, since 2008. Genifuels was founded in 2006 as a producer of equipment to produce renewable fuels from wet organic materials. Since crude oil and natural gas are the products of the system, the question remains is the cost of the products less than the price of crude oil of about $96/bbl (West Texas Intermediate) at the beginning of 2014. Their efforts to produce crude oil directly from algae are reminiscent of those of KIOR who are seeking to do so from waste cellulose as described in Chapter 3.

Several other firms are also pursuing the direct conversion of algae to crude oil. Among them are Origin Oil of Houston, TX, and Applied Research Association of Albuquerque, NM. Other than the fact that Origin Oil is codeveloping an integrated system with DOE's Idaho National Laboratory, little detail has been published about their technology. Applied Research is using an aqueous system that has a catalytic reactor as the centerpiece [52]. This employs high pressure (15−20 atm) and somewhat moderate temperatures of 240−450°C along with water and catalysts. In 2012, Applied Research joined forces with Chevron Lummus Global, the National Research Council of Canada, the U.S. Air Force Research Laboratory, and Arisoma Biosciences to develop a

100% drop-in jet fuel. They expect the fuel will meet all ASTM requirements and avoid the blending quality issues surrounding other jet fuel candidates.

Solazyme of South San Francisco, CA, uses straightforward chemistry in which triglycerides from algal oils are transesterified with methanol to form the corresponding methyl esters. The so-produced fuel is useful as a drop-in fuel without modification in any ratio with conventional diesel fuel for existing engines. Solazyme products are combined with petroleum diesel as an 80:20 bio/petroleum diesel blend marketed as Biodiesel B20. The blended fuel was sold successfully on a test market retail basis from several California gas stations at the end of 2012. A number of products can be produced by combining their algal oil with one or more of those in Table 6.2. In contrast to the use of open ponds or a PBR, Solazyme grows its algae in the dark inside large stainless steel containers. Their proprietary algae is heterotrophic and consumes a variety of carbohydrates including sucrose. Solazymes proprietary algae are capable of producing as much as 75% of their dry weight as oil. They expect to produce as much as 20,000 metric tons of biodiesel at their Clinton, Iowa, plant in 2015. This is an Archer-Daniels Midland facility. They are also planning a 100,000 metric ton facility in Brazil by year end 2014.

A dramatically different system from the type being pursued by Solazyme is that of Algenol Corp. of Bonita Springs, FL. Their process is entitled *Direct To Ethanol* and does not rely on algae recovery and processing but on the natural ability of Cyanobacteria to produce ethanol in algae. In Algenol's approach, they enhance the endogenous production of ethanol directly rather than engage on its production as part of routine cell maintenance. A central feature of their system is a proprietary plastic film bioreactor that facilitates ethanol production and recovery. Each bioreactor has ports for recovery of ethanol and waste biomass and also ports for the introduction of carbon dioxide and nutrients. Ethanol is produced at the rate of 10,400 gallons/acre by using sunlight, carbon dioxide, and salt water. They claim a production cost of about $1.18 per gallon and plan to market ethanol for as little as 75 cents per gallon below its current market price. Industrial grade carbon dioxide off-gases from refineries and power plants can be used. In their system, one metric ton of carbon dioxide yields 160 gallons of ethanol and 2 gallons of fresh water.

This is equivalent to 0.48 lbs of ethanol per lb of carbon dioxide. Reliance Industries of India has reportedly invested $93.5 million in the development.

Algenol uses engineered blue green algae (Cyanobacteria) that converts the pyruvic acid produced in photosynthesis to ethanol and carbon dioxide. Cyanobacteria grow faster than plants and also absorb sunlight more efficiently. Accordingly, they do not require plant matter to synthesize either isobutanol (see Chapter 5) or ethanol. Algenol claims to have about 2300 strains of algae that they have collected globally and screened for ethanol production. In addition to ethanol, Algenol also plans to forward integrate into jet and diesel production.

Although ethanol production from carbon dioxide, cyanobacteria, and water looks simplistic in Algenol's hands, its central feature of photosynthesis is considerably more complex. The series of reactions that occur in photosynthesis are illustrated in Figure 6.2 in which the formation of 2-phosphoenolpyruvic acid (PEPA) is a key intermediate for ethanol formation. The carbon reduction cycle on which it is based comprises the carboxylation of ribulose-1,3-diphosphate with carbon dioxide to yield a highly labile beta-keto acid. This is subsequently converted within the cell as illustrated in Figure 6.3.

A similar system to Algenol's is that of Joule Unlimited (Cambridge, MA). Their system [56] employs proprietary genetically modified bacteria that can produce ethanol or diesel in a *Solar Converter*. The later is also referred to as a *Helioculture Platform* and resembles a series of solar panels. By year end 2013, Joule had received at least six US patents for their system which they claim can produce renewable fuels at a rate 10 times faster than processes based on cellulose. They commissioned their first demonstration plant (25 million gallon capacity) at Hobbs, NM, at the close of 2012. The product, Joule Sunflow-E, is designated for the ethanol market. A second product, Joule Sunflow-D, is under development as a diesel fuel. They claim that this can be readily blended with 50% petroleum diesel. Like Algenol, Joule employs genetically modified Cyanobacteria that directly convert sunlight and carbon dioxide to fuel.

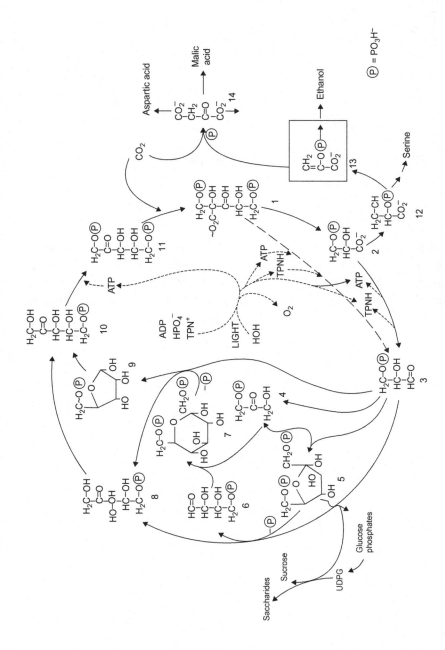

Figure 6.2 Overall photosynthesis pathway [53].

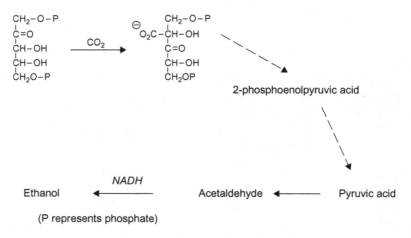

(P represents phosphate)

Figure 6.3 Photosynthetic pathway to ethanol.

With all the activity on the use of algae, ExxonMobil's decision to cut back on its 4-year algae program with Synthetic Genomics, Inc. is mystifying. The partners who have spent more than $100 million have reportedly decided to focus on basic genomics instead of process development.

CHAPTER 7

Diesel Fuel

As a fuel for internal combustion engines, diesel ranks second only to gasoline. In 2012, total US demand for diesel fuel was 57 billion gallons, which was less than half that of motor gasoline. In early 2014, its \$3.95/gallon price was higher than that of regular gasoline which sold at \$3.65/gallon. Unlike gasoline, diesel fuel has a large number of end uses as shown in Table 7.1.

The development of the diesel engine and biofuels is closely linked historically. Rudolf Diesel had a revolutionary theory in the nineteenth century. He envisioned an engine in which air was compressed to such a degree that there is an extreme rise in temperature. When fuel was injected into the piston where the temperature rise had taken place, it exploded and forced the piston down enabling it to drive the wheels of a vehicle. He was attempting to surpass the performance of a steam engine which was then the state of the art as a power source. His first diesel engine, which was introduced in 1897, was powered by vegetable oil, peanut oil in this instance, and had an extraordinary efficiency of 75%. As a result, diesel engines and biofuels debuted simultaneously, and diesel compression engines were powered by vegetable-based oils until the 1920s when they were supplanted by petroleum-derived fuels from the then fast-growing refining industry.

Rudolf Diesel literally disappeared on the eve of World War I in 1913. He had had a falling out politically with the German government with whose policies he disagreed and who wanted to use his diesel engine for their navy. He was traveling to Great Britain to offer his technology when he inexplicably disappeared over the side of his ship while crossing the English Channel.

The 1920s brought a new injection pump design that allowed fuel to be metered as it entered the diesel engine. This enabled the elimination of pressurized air in an accompanying tank. The diesel engine was now small enough to be conveniently used in motor vehicles. Mercedes Benz

Table 7.1 Diesel Fuel End Uses	
End Use	Percent of Total
On-highway motor vehicles	64
Off-highway and farm applications	9
Residential	6
Railroad locomotives	5
Ships	3
Miscellaneous	13
Total	100

introduced the first such vehicle in 1936. Concurrently, diesel engines were altered in design so that they could use lower viscosity fossil fuels in place of such biomass-derived fuels as vegetable oils.

A dramatic surge in diesel powered automobiles occurred after the 1978 Iranian Oil Embargo when the public looked to diesel engines for improved fuel economy. Sales of diesel powered vehicles surged dramatically such that they accounted for 85% of Peugeots that were sold, 70% of the Mercedes Benz market, and 50% of the Volkswagens. By 2000, when the furor over oil supplies had subsided, diesel sales fell dramatically, such as, for example, to where they were only 8% of Mercedes Benz cars sold. Only slightly more than 10 years later, interest in diesel has reawakened because of environmental and energy conservation issues.

When diesel fuel is atomized and intimately mixed with hot compressed air, a number of ignition points are created throughout the cylinder. This process enables early and uniform ignition. Ignition is not instantaneous, however, and depends on the hydrocarbon composition of the fuel. The ignition time is shorter for linear paraffinic fuels than for those containing cycloaliphatic, olefinic, and aromatics. Long delays in ignition cause excessively rapid rises in pressure after combustion and result in rough engine operation. Such delays also allow time for certain unwanted chemical reactions to occur, and the products of those unwanted reactions burn very rapidly leading to a very unwanted rise in pressure. Diesel fuels are accordingly defined by a cetane number that is based on its paraffinic content. The cetane number therefore measures the fuel's ignition delay characteristics.

Besides causing rough engine operation, long ignition delays cause misfiring and uneven combustion in a cold engine. The result is a loss in power and smoke in the engine exhaust. Rating the cetane quality of a diesel fuel is equivalent to determining the octane rating of gasoline. A normal C_{16} paraffin, such as n-hexadecane, has a very high ignition quality and represents 100 on the cetane scale. This compound is defined as cetane *per se*. On the other hand, 2,2,4,4,6,8,8-heptamethyl-nonane (HMN), a highly branched saturated hydrocarbon, is assigned a cetane number of 15 and is the low quality reference fuel. Blends of the two hydrocarbons represent intermediate ignition qualities, and their cetane number is calculated from the percentages of each in the blend. Consequently, a reference blend of 35% cetane and 65% HMN has a cetane number of 45. Accordingly, higher cetane numbers accrue to molecules inversely to what leads to higher octane numbers in gasoline. Aromatics and highly branched aliphatics are attractive for higher octane gasoline.

Cetane numbers are determined in a manner similar to the technique used for octane determinations by testing the fuel in a one cylinder engine run at 900 rpm. Octane numbers are reported as research (RON) and motor octane (MON) with the actual number on the gasoline pump in the United States being the average of the two. RON is the result obtained from a one cylinder engine run at 600 rpm and MON is determined at 900 rpm. There is no direct link between RON and MON except that MON is generally 8−10 points higher, nor is there any direct link between cetane and octane.

The density of petroleum diesel at 6.9 lbs/gal is about 12% greater than ethanol-free gasoline (6.2 lbs/gal). Because of its greater density, petroleum diesel has greater volumetric energy than regular gasoline (128,700 BTUs/gal vs. 115,500 BTUs/gal). Carbon dioxide emissions from diesel are only slightly lower than that of regular gasoline. Petroleum diesel is obtained by fractional distillation of crude oil at 180−360°C. It is generally composed of 75% saturated hydrocarbons (both straight chain and branched) and 25% aromatics. Carbon content is typically 8−20 per molecule.

The price of diesel generally rises in the colder months as demand for heating oil also rises since diesel is obtained from the same crude oil source in much the same way. Changes in fuel quality regulations to remove sulfur also contribute to diesel's higher price. The introduction of ultra-low sulfur diesel regulations to accommodate minimization of greenhouse gases has further complicated the production of petroleum diesel and added to its cost. As of 2006, nearly all petroleum diesel produced in North America is ultra-low sulfur diesel. The maximum sulfur content in US diesel must not exceed 15 ppm. This is a decrease from the 500 ppm maximum that had been permitted to accommodate the emission control systems of cars produced since 2007. This same standard will be applied in 2014 and 2015 to diesel fuel used in locomotives and ships. Diesel produced from renewable sources has a decided cost of production advantage in this respect since sulfur is either absent or very low in such systems. If present at all, sulfur in biodiesel results from impurities contracted in the process of diesel production. The presence of sulfur in petroleum diesel is endogenous and results from the crude oil from which it is fractionally distilled.

A breakdown of the factors inherent in the cost structure of petroleum diesel compared to gasoline is illustrated in Table 7.2. Distribution and marketing costs for diesel are higher while that for crude oil is decidedly less. Most likely a similar cost structure accrues to biodiesel. The only uncertainty is the cost of vegetable oils compared to crude oil. The $100 per barrel market price for crude oil in early 2014 is equivalent to about $0.31/lb. The prices of soybean or corn oil, for example, at that time are slightly higher at $0.42/lb. If mixed unrefined vegetable oils were employed, it is probable that their price might be about the same as that of crude oil.

A number of processes are under development for the production biodiesel. Among them are those processes described in the chapters on cellulose gasification and fermentation (see Chapter 5) and algae (see Chapter 6). The most commonly used system to date is the so-called FAME (fatty acid methyl ester) technique in which vegetable oils are

Table 7.2 Price Structure of Petroleum Diesel Versus Gasoline [57]		
	(Percent of Total)	
	Diesel	Regular Gasoline
Taxes	12	12
Distribution and marketing	14	6
Refining	15	12
Crude oil	59	70
Total	100	100

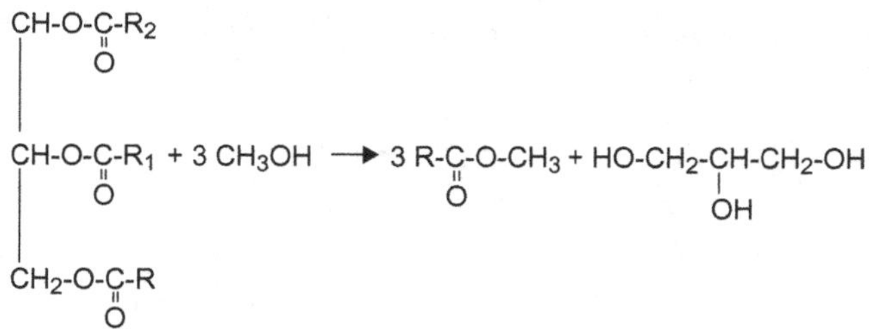

$$\begin{array}{l} CH\text{-}O\text{-}\underset{\underset{O}{\|}}{C}\text{-}R_2 \\[2em] CH\text{-}O\text{-}\underset{\underset{O}{\|}}{C}\text{-}R_1 \ + \ 3\ CH_3OH \longrightarrow 3\ R\text{-}\underset{\underset{O}{\|}}{C}\text{-}O\text{-}CH_3 + HO\text{-}CH_2\text{-}\underset{\underset{OH}{|}}{CH}\text{-}CH_2\text{-}OH \\[2em] CH_2\text{-}O\text{-}\underset{\underset{O}{\|}}{C}\text{-}R \end{array}$$

Figure 7.1 FAME reaction.

transesterified with methanol. Virtually any lipid can be used in that system and ethanol is an optional reactant for methanol. The overall reaction can be summarized as shown in Figure 7.1, where R, R_1, and R_2 can be the same or different hydrocarbon segments.

Byproduced glycerol can be recovered as a byproduct credit. Virtually any lipid can be employed, such as rapeseed, coconut, or soybean oil. The transesterification is usually catalyzed by NaOH or KOH. FAME products have a lower energy content than petrodiesel because of their oxygen content and they might have higher nitric oxide emissions, possibly even higher than the legal limit. An alternate system [58] employs NaOH or KOH-free esterification and proceeds under supercritical conditions at high temperatures. In this technique, the methanol and lipids are together in a single phase and the reaction is much faster. This reaction variation tolerates the presence of water and any free fatty acids are converted to esters instead of potassium or sodium salts. The reaction commonly proceeds in a tubular reactor at 350°C and pressures of 10–19 Mpa. The methanol to lipid ratio is 42 and conversions of 95% are obtained. As expected, FAME type diesels have very low sulfur content and also exhibit 30% lower carbon monoxide emissions. Most importantly, exhaust emissions of total hydrocarbons are up to 93% lower than petrodiesel. FAME type diesels have a series of drawbacks, however, such as corrosion of fuel injection ports, low-pressure fuel systems, and pump seizures because of higher fuel viscosity.

UOP (Des Plaines, IL), a subsidiary of Honeywell, has introduced a process for diesel from vegetable oil in which transesterification is eliminated [59]. Entitled Ecofining, the UOP process entails direct catalytic hydrogenation of the lipid triglycerides to produce paraffinic hydrocarbons. In their system (Figure 7.2), the lipids are combined with 2.5–3.8% recycled hydrogen and sent to an adiabatic multistage reactor.

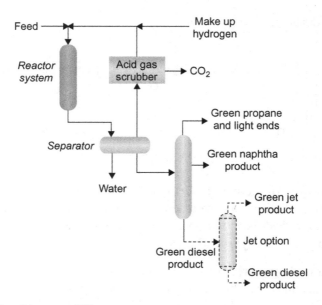

Figure 7.2 UOP ecofining process [59].

The lipid feedstock is completely deoxygenated and converted to saturated hydrocarbons. Conversion and selectivity is 100% and the products are in the diesel boiling point range. The effluent from the reactor
is separated at reactor pressure to remove carbon dioxide, water, and
such low molecular weight hydrocarbons as propane. Selectivities to the
products are propane/naphtha/diesel: 2−4/1−10/88−98 v/v. The longer
chain paraffins are combined with additional hydrogen and sent to a
hydroisomerization unit (not shown) where the diesel is recovered. UOP
refers to the product as *Green Diesel* and claims that the process is readily applicable to algal lipids. A comparison of this product with diesels
produced by others is illustrated in Table 7.3.

Eni will be retrofitting its Venice, Italy, refinery to produce UOPs
Green Diesel. The plant will produce 100 million gallons annually
beginning in 2014. The parties claim that Green Diesel can be used on
a drop-in basis without blending.

A system somewhat similar to UOPs is that of Neste Oil. The process, entitled NExBTL, is based on direct catalytic hydrogenation of
plant oils to their corresponding alkanes. As in the UOP process, the
glycerol segment of the triglycerides is converted directly to propane so
there is no need for glycerol separation from the reaction products.

Table 7.3 Properties of Diesel Products [59]				
	Ultra-Low Sulfur	Green Diesel	FAME	FT Diesel
Oxygen, %	0	0	11	0
Specific gravity	0.84	0.78	0.88	0.77
Sulfur, ppm	<10	<1	<1	<1
Heating val, MJ/kg	43	44	38	44
Cloud pt., °C	−5	−20 to +20	−5 to +15	n.a.
g CO_2 eq/MJ	n.a.	50	n.a.	110
Cetane	40	70−90	50−65	>75

Neste appears to be ahead of Eni, UOP's licensee, with two plants brought on stream in 2007 and 2008, each with a diesel capacity of 0.17 million tons. Two larger such plants of 0.8 million tons each were brought on stream in 2010 and 2011 in Singapore and Rotterdam, respectively. Renewable diesel has been profitable for Neste with more than $165 million in total earnings in 2013. The company is further diversifying its renewable fuels portfolio by developing systems to produce microbial oils from carbohydrate waste using yeast and fungi. Pilot plant construction for the latter began in 2012. They intend to use this newer development to furnish the oil feedstock for the production of diesel by the NExBTL process.

According to ExxonMobil [60], diesel will surpass gasoline as the number one transportation fuel worldwide by 2020 and will continue to increase its market share through 2040. The shift to diesel away from motor gasoline is being driven by improving economy for light duty vehicles and the growth in commercial transportation activity. Diesel demand accounts for 70% of the growth for all transportation fuels through 2040. In contract to this, ExxonMobil expects gasoline demand to be relatively flat despite a doubling in the number of personal vehicles from 800 million to more than 1.6 billion. An indication of the growth in diesel usage, especially biodiesel, became apparent at the end of 2013 when demand reached its highest level ever of 35 million gallons for December alone.

ExxonMobil's forecast is echoed by a report from J.D. Power Associates that diesel's near term share of the US market for automobiles and light trucks will be more than triple to 10%. Higher prices for diesel ($3.95/gal) versus gasoline at $3.65/gallon are offset by diesel's superior mileage of 35 mpg (highway) versus 26 mpg (highway) for gasoline. Diesel also leads gasoline for mileage in city driving of 26 mpg compared to 19 mpg.

As noted in Chapter 4, FT product distribution is ideally suited for diesel production. A negative feature of FT technology is its broad product distribution. This can be controlled within limits by reaction temperature and the degree of synthesis gas conversion to products as shown in Figure 4.5. Diesel molecular weight largely falls within the range of middle distillates.

Despite its long commercial history, only five FT plants are operating worldwide where they largely take advantage of plentiful, low-cost natural gas. Shell Oil briefly considered a plant in Louisiana and abandoned its plans at year end 2013 because of high fixed capital investment. The investment cost for the 140,000 barrel per day plant would have been >$12.5 billion [64]. In an earlier project, the investment cost for Shell's FT plant in Qatar was originally estimated to be $5 billion, but the final cost for the completed project was about $20 billion. The view that FT technology could possibly change with the introduction of the innovation by Velocys was discussed briefly in Chapter 4. Their microchannel reactors may enable the production of FT diesel to the extent that a plant 1500 barrels per day in size may be as economical as a conventional one that is 30,000 barrels per day. Accordingly, a Velocys plant offers the prospect of much lower fixed capital investment, such as discouraged Shell Oil from moving forward in Louisiana. Ostensibly, the Velocys reactors are compact and have tube sizes in the millimeter range as opposed to those in the centimeter range that is typical of conventional plants. When the microchannel reactors are used in conjunction with the cobalt catalysts developed by Oxford Catalysts, selectivities to products of 87% at more than 90% conversion are possible. In contrast, conventional FT facilities typically have conversion efficiencies of only 50% per pass and are designed to operate at a minimum of 5000 barrels per day.

Velocys is a 2001 spin-off from Battelle Memorial Research and was originally named Oxford Catalysts. The latter was formed in 2004 and is a product of Oxford University's Wolfson Catalyst Center. An order has been placed for the first module of a 2800 barrel per day FT plant by Pinto Energy in Ohio. The plant is ostensibly natural gas based. Other active projects are a 1000 barrel/day liquid fuels plant at

Calumet in the United States and an \$8 million order from Ventech. The microchannel technology [63] was developed at Battelle Institute and the first patent was issued in 1994. Oxford Catalyst and Velocys merged in 2008. The success of the technology could generate a resurgence of FT chemistry from either waste cellulose or natural gas. The latter has a cost advantage in terms of feedstock that can be delivered to a plant site by pipeline. Collection and shipment of waste cellulose could be meaningfully greater.

CHAPTER 8

Natural Gas

8.1 INDUSTRY BACKGROUND

Although natural gas is a nonrenewable resource, it is included for discussion because its sudden growth in production will have an impact on the development and use of renewable fuels. Natural gas supplies fall into two categories based on their origins: (1) free unassociated gas not in contact with crude oil and (2) gas that is associated (dissolved) in crude oil. Added to these sources is the recent advent of hydraulic fracking that is responsible for recovery of very large quantities of gas from shale, originally in the United States, and ultimately worldwide. Unprocessed natural gas is commonly associated with liquids such as ethane, propane, and butane, each of which has a higher BTU value than natural gas itself and accords a higher value if not separated when natural gas is sold for heating purposes. These products are commonly referred to as natural gas liquids (NGLs) and are attractive for sale to the chemical market, primarily for production of ethylene. These are discussed briefly toward the end of this chapter. The principal component of natural gas is methane, and this term and natural gas are frequently used interchangeably. To obtain a marketable product, natural gas recovered at the wellhead must be processed to remove water and such deleterious products as hydrogen sulfide.

Natural gas is a very desirable fuel for internal combustion engines. On the basis of cost per joule of heat energy, it can be the least expensive fuel and its high octane rating enables the design of very efficient high compression engines. The properties of natural gas and NGLs are illustrated in Table 8.1.

Natural gas is often the least expensive engine fuel in a given location. In its dry sulfur-free state in which it is available from pipelines, it is a very clean fuel that burns cleanly in engines, and it produces very little engine deposits or corrosion.

Table 8.1 Properties of Natural Gas and NGLs			
	Heating Value[a]	Motor Octane No.	Vol. Pct. In Nat. Gas
Methane	913−1013	130	87.1
Ethane	1641−1793	103	3.4
Propane	2385−2540	99.6	1.0
n-butane	3113−3370	91.6	0.5
[a]*BTUs per cubic foot.*			

Hydraulic fracking, commonly referred to as *fracking*, is a technique in which water-containing chemicals and sand are injected into a wellbore under pressure to create small fractures of less than 1 mm in size and through which gas or oil can migrate to the wellhead. The pressure is then removed from the well and small grains of sand or aluminum oxide are used to keep these tiny fractures open once the composition of the rock formation has stabilized. The composition of the chemicals that are used varies from well to well. Fracking is applied usually but once in the life of a well and quickly enhances product removal and productivity. As noted in Chapter 1, the first commercial application of fracking was in the 1990s although the first experimental application was much earlier in 1947. By 2010, about 60% of all new gas and oil wells employed fracking [16]. Shale accounts for 32% of all new natural gas resources. As of 2012, there were more than 2.5 million applications of fracking with more than one million of them in the United States.

Pan American Petroleum made the first large-scale use of fracking in Stephens, OK, and they were followed by its use in thousands of wells in the San Juan Basin and other sites in western United States. A major segment of such growth is in the Bakken region of North Dakota that now ranks second to Texas in natural gas and oil production. The Bakken shale formation is located in the region bordering North Dakota and Montana and is estimated to contain 43 billion barrels of oil and gas. This is the largest find in US history. Production in this region has outpaced North Dakota's pipeline capacity and processing facilities. As a result, the amount of nonmarketed natural gas has grown to 3.1 billion cubic feet per day (BCFD).

Another notable area for growth in the United States is the Marcellus region of Pennsylvania and West Virginia. Production in

Table 8.2 US Natural Gas Foreign Trade (TCF)		
	Exports	Imports
2008	0.96	3.98
2009	1.07	3.75
2010	1.14	3.74
2011	1.51	3.47
2012	1.62	3.14
2013	1.57	2.88

this region exceeded 13 BCFD in December 2013 compared to less than only 2 BCFD in 3 years earlier and will account for 18% of total US natural gas production. The American Gas Association estimates US reserves to be about 300 billion cubic feet, a record with 100 years of supply at 2013 rates of consumption of 22–23 trillion cubic feet (TCF) per year. Accordingly, natural gas imports into the United States have been declining while exports are static as illustrated in Table 8.2. The US Energy Information Administration [63] expects imports from Canada to fall by 30% from 3.0 to 2.1 TCF by 2040 as more domestic demand is met by production. On the contrary, exports to Mexico by pipeline are projected to increase 6% annually from 0.6 TCF in 2012 to 3.1 TCF in 2040.

As would be expected, fracking has quickly spread to western Canada, both the offshore and onshore gas fields in the Netherlands and the UK sector of the North Sea. For much of Europe, however, fracking has been out of reach because of environmental concerns [64]. The governments of France and Germany are holding back on fracking permits because of public concerns about safety and the environment. Germany, for example, would otherwise have had at least a 10-year supply of shale gas if it used fracking. Similar concerns do exist in the United States and are the subject of ongoing debate at the state and local level. There is similar debate in the United Kingdom, and there is government support for fracking.

US demand for natural gas by end use is summarized in Table 8.3. It has grown at the rate of 2.1% annually from 21.4 to 23.9 TCF per year from 2008 to 2013. Electric power generation followed by industrial applications are the primary end uses. Although use in vehicles is only a small fraction of the total demand at less than 1%, it has been growing at a 5% annual pace since 2008. This is more than double the

Table 8.3 US Natural Gas Consumption [63] (TCF)			
	2008	2010	2013
Electric power	6.67	7.39	8.15
Industrial	6.67	6.83	7.46
Residential	4.89	4.78	4.94
Commercial	3.15	3.10	3.29
Vehicle fuel	0.026	0.029	0.033
Total	21.41	22.13	23.87

rate of natural gas as a whole and even greater than the 4% annual rate for use in power plants. Consumption for power plant generation is driven by displacement of coal as a pollutant for such plants.

As natural gas production has grown, wellhead prices have sharply fallen from $7.97 to $2.66 per thousand cubic feet (MSCF) from 2008 to 2012. Similarly, the natural gas price for vehicles fell nearly 30% from $11.75 to $8.03/MSCF in the same time period. Increased production volume in the United States combined with the drop in prices and the environmentally friendly character of the gas has made it a potentially attractive candidate as a motor fuel, at least for trucks and buses, in the United States. In contrast, natural gas prices are considerably higher in Asia and South America as shown in Figure 8.1. US spot natural gas prices are less than $3 per million BTUs (MMBTU) compared to $9/MMBTU for spot gas in Europe and $11/MMBTU on European oil-linked contracts and as $11/MMBTU in Asia. The United States has been described [65] as an island in gas terms.

The large US volume of natural gas reserves has led to agitation by producers for the construction of more liquefied natural gas (LNG) terminals and for Federal approval to pursue foreign markets. Construction began for the first west coast terminal in late April, and in all, seven LNG facilities have been approved. Total export capacity is 10 BCFD. A coalition of US manufacturers, among them Dow Chemical and Huntsman Corp., believes that large volume exports of LNG will raise domestic prices while oil and gas companies argue that exports will have little effect on domestic prices. Critics argue that with the newly found surplus in US natural gas, the readiness to export it appears to be a grievous error for the sake of immediate profit.

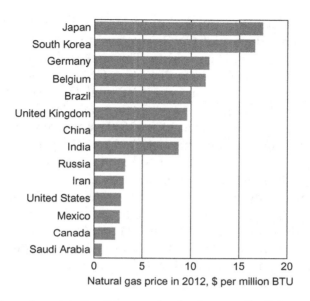

*Figure 8.1 **Global natural gas prices.*** *Note: Prices generally reflect domestic wellhead/hub prices or prices for gas imported via pipeline. Some nations, such as Japan and South Korea, have higher prices because they import LNG.* American Chemistry Council.

Although natural gas is environmentally friendly from a combustion point of view, it should be borne in mind that it is *methane* nevertheless, a powerful greenhouse gas that is 72 times more potent than carbon dioxide. This is true for the first 20 years that the gas is in the atmosphere and it is 20 times more potent over 100 years. If it leaks to the atmosphere as opposed to being burned, it quickly loses its greenhouse gas advantage.

8.2 VEHICLE FUEL

There are three types of natural gas vehicles, those that are dedicated, those that have separate fuel systems that enable them to run either on natural gas or gasoline, and those that are dual feed. These last vehicles are traditionally limited to heavy-duty applications in that they run on natural gas and use diesel for ignition assistance. Light-duty vehicles typically operate in a bifuel mode and heavy-duty vehicles operate in either dual feed or dedicated modes.

Natural gas vehicles use compressed natural gas CNG and LNG as fuel with the former being the most common. The form of natural gas that is chosen depends on the range in which the vehicle operates. Since the energy density of CNG is greater than for LNG, more fuel can be stored on board the vehicle. This makes it more attractive for trucks that must travel greater distances. CNG vehicles have been in use for decades and account for 2% of total US demand for vehicle fuel [67]. Most CNG is employed in fleet vehicles such as taxis and buses. CNG fuel is held under 3000–3600 psi pressure and the tank is consequently more massive, heavier, and more expensive than a conventional gasoline or diesel tank. LNG is the same methane-containing material as CNG but it has been condensed to a liquid under atmospheric pressure by cooling to $-162°C$. As a result, its volume is reduced to 1/600 that of gaseous methane, and its energy density is therefore 2.4 times greater than CNG or 60% that of diesel fuel. This makes LNG cost effective to transport over long distances where there is no pipeline. Specially designed cryogenic tankers and onshore terminals are used for shipping LNG.

Existing gasoline powered vehicles can be converted to CNG or LNG. Diesel engines for heavy trucks and buses can be converted to CNG by the addition of new heads that contain special spark ignition systems, or they can be run on a blend of diesel and natural gas. In this last instance, the primary fuel is natural gas with a small amount of natural gas being added as an ignition source.

Until recently, the Honda Civic GX was the only natural gas passenger automobile commercially available in the United States. This has now been joined by Ford, GM, and RAM. Ford offers a "bifuel prep kit" as a factory option and then lets its clients select an authorized party for installation of natural gas equipment. RAM is the only US pickup truck manufacturer that offers a truly factory installed bifuel system.

LPG (liquefied petroleum gas) can also be used as vehicle fuel but it is a totally different product. It is not natural gas at all, although its components are cryogenically recovered from natural gas. Such components are ethane, propane, and butane and are not the subject of this chapter. LPG is frequently referred to as autogas and possesses a number of the advantages and disadvantages of LPG and CNG. Although lower in cost, LPG's greatest drawback is safety because it is heavier than air with a tendency to collect in low spots. This makes LPG somewhat hazardous to use. LPG also has lower energy density than either gasoline or diesel and therefore offers poorer mileage.

On the positive side LPG has a high research octane rating of 102−108 and burns more cleanly than either diesel or gasoline. LPG has literally many uses in the home, in industry, farms, and businesses wherever heat, light, or power are required and natural gas is otherwise unavailable. A common home use is lighting gas grills and fireplaces. It is used as an aerosol propellant and a refrigerant in place of chloro-fluorocarbons that have been banned from use. The recovery of LPG from natural gas is growing rapidly and set a record in 2013 when its demand increased to 1.8 million barrels per day at year end.

CNG vehicles are nearly 25% more expensive than their conventional gasoline counterparts and nearly 10% more than hybrid vehicles. According to ExxonMobil [66], a CNG powered passenger car available in the United States costs about $5600 more than a similarly equipped gasoline model, and a CNG powered 18 wheeler van costs an additional $60,000 more in 2012. CNG vehicles also have an issue with highway infrastructure since an entirely new pipeline network and refueling stations would be needed to service them.

In 2013, the United States had a fleet of 112,000 CNG vehicles, most of them being buses [68]. Another 150,000 vehicles ran on LPG and about 4000 ran on LPG. There are about 18 million natural gas vehicles worldwide as noted in Table 8.4.

Natural gas vehicles have a very small share of the market in the United States, but there are indications that this may change with the burgeoning availability of natural gas (Figure 8.2). The Ford Motor

Table 8.4 Major Countries by Natural Gas Vehicles (2013)	
	Million Vehicles
Iran	3.5
Pakistan	2.7
Argentina	2.3
Brazil	1.8
China	1.6
India	1.5
All other	4.7
Total	18.1

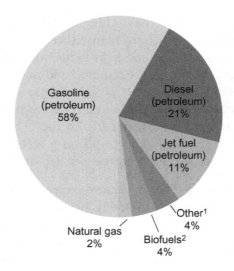

Figure 8.2 Fuel used for US transportation, 2010.

Company, for example, reported the sale of 11,600 such vehicles in 2012, triple the number that it sold only 2 years earlier.

There are projections by several consulting companies and government agencies who see natural gas as a significant new transportation fuel in the next decade to the extent that it may challenge crude oil for the automotive market. The Paris-based International Energy Agency, for example, expects natural gas consumption worldwide for motor vehicles to reach 3.6 TCF by 2018. Similarly, Boulder, CO, based Navigant Research [69], the consulting company, expected the sale of 930,000 natural gas vehicles in 2012. The bulk of these sales were CNG powered trucks. They also offer their view of 20,233 CNG refueling stations worldwide with nearly half of them being in the United States. The corresponding forecast for LNG refueling stations is 117.

Two main issues govern increased usage of natural gas in cars and trucks: (1) the lower price of natural gas at $1.47 per gallon versus that of diesel at $3.97 per gallon in 2014 and (2) the lower emissions from natural gas engines. The CNG and LNG used in buses and trucks in the United States are priced well below those of most countries and commercial fleets can benefit from a total cost of ownership than that which is less than for diesel vehicles. The demand for natural gas powered trucks and buses is uneven on a regional basis, and North America and the Pacific Rim countries stand out as regions of strong growth. In North America where natural gas prices are low, the growth of the vehicle market has gotten ahead of refueling station development. In the Pacific Rim, developing markets look to natural gas as a solution to environmental problems. Navigant Research [69] believes that the total number of light-duty natural gas vehicles world-wide will reach 25 million by 2022.

In contrast to the environmental appeal seen in natural gas vehicles, some parties believe it may actually hurt the climate. In a study done by Stanford University, the authors conclude that unavoidable methane leaks that occur during drilling and natural gas processing offset the benefit the gas by lower carbon dioxide emissions from burning. They conclude that even the use of natural gas in passenger cars is borderline in terms of damage to the environment. The Stanford conclusions [70] were not based on original research but on a review of more than 200 studies done by others. An industry group, Natural Gas Vehicles for America, strongly disagrees and cites a report by the California Energy Commission that calculated a 22% reduction in greenhouse effects on a "well to wheel" basis that can be had by the use of natural gas. Their calculations included extraction and distribution.

Although natural gas is making large inroads in the truck and bus market, the most attractive potential opportunity is for passenger automobiles. Public refueling stations for automobiles are few and far between, and there is the question of consumer psychology. How does one convince drivers that it is safe to put natural gas in their cars? The big issue is the fuel tank. Gasoline and natural gas engines are similar, but since natural gas must be stored under pressure (3600 psi) for use as CNG, the tanks must be stronger, heavier and larger, and this drives up the price of the car. As already noted, a Honda Civic GX costs about $5200 more than a comparable car powered by gasoline. It is even about $3600 more expensive than the gasoline/electric hybrid Civic. At a $5500 premium and a fuel price of only $1.40 per gallon, it could take an owner about 9 years to break even. Barring unlikely legislation by Congress that would compel a switch to CNG vehicles, an effort is required to reduce the cost of the fuel tank. One such effort in progress is that of 3M Corp. and Chesapeake Energy Corp. to develop a CNG tank that uses a plastic lining wrapped in a carbon composite. Such a tank might be 10–20% lighter, they argue, and would have 10–20% more capacity than current CNG tanks. Another effort at the University of Missouri would store natural gas at much lower pressure by keeping it in a material made of corn-cobs fabricated into charcoal briquettes.

A number of companies have begun to set up refueling stations for fleet vehicles. Some such stations are in prominent public locations and advocates hope they will spark interest in natural gas vehicles. Apache Corp., for example, built a CNG refueling station at Houston's International Airport to service a small fleet of CNG parking shuttles operated by the City of Houston. The big barrier to setting up refueling stations nationwide is its cost. The average cost for a combined gasoline station and convenience store in the United States was about $2.3 million in 2010. The addition of a compressor and CNG storage tanks would add $500,000 to this cost, and this is on the assumption that the station would back up to a natural gas distribution pipeline. An obvious compromise would be to install CNG facilities at existing gasoline service stations. Although there would be a capital investment but it would largely be confined to compressors and CNG storage tanks. Proximity to a natural gas pipeline would still be essential.

An illustration of possible natural gas penetration into the passenger vehicle market can be illustrated by consideration of diesel penetration into the market for heavy-duty trucks [71]. Such trucks weigh more than 33,000 pounds and are the so-called tractor trailers. Their average age is about 7 years and it takes about 14 years for the entire fleet to turn over. The average age for a passenger vehicle is even longer at 9 years which means that about 18 years are required for the complete turnover of a fleet. In practice, new technologies do not have a linear growth rate for market inroads and generally follow an S-shaped growth curve. Support for this view is found in the historic example how heavy-duty trucks shifted from gasoline to diesel in 1950. About 90% of such vehicles were fueled by gasoline at that time. Diesel vehicles captured a small but increasing share of the market over the decades that followed to the extent that their market share rose to about 90% by 1970. Diesel continued to increase its share in the 1990s when it had completely displaced gasoline. It only took about 25 years to become an overnight replacement! Diesel faced considerable challenges to a new market entrant in competing against an entrenched incumbent in gasoline. Diesel engines had higher upfront costs, added more weight to the vehicles, and required a new refueling infrastructure for support. Diesel succeeded despite these obstacles that bear a great similarity to those for natural gas vehicles. In this instance, patience is a virtue.

An alternative to using natural gas directly in vehicles is to first convert it to a conventional fuel. Such an approach avoids all the issues in the cost of a new infrastructure to support it. Some parties have followed this route by first converting natural gas to a synthesis gas (Chapter 4). In a new approach taken by Siluria Inc. [72,73] of San Francisco, methane is directly oxidative coupled to ethylene that is then catalytically coupled to liquid fuels. This system appears to avoid the high capital investment inherent in Fischer Tropsch but they are not the first party to pursue oxidative coupling. Others have pursued this for ethylene production for the chemical market and have largely abandoned it. Insufficient information is available from their patent applications to evaluate the probable success of its development.

Methanol is a potentially attractive alternative to natural gas as a fuel since it can be easily shipped as such and doesn't require prior conversion to LNG. Currently, however, methanol is produced from natural gas by first being converted to a synthesis gas (Chapter 4). This is an intrinsic feature of the ExxonMobil methanol to gasoline process. An alternative effort to that of Siluria in 2012 is by a group at Dow Chemical and Cardiff University in Wales who reported the direct conversion of methane to methanol [75] in greater than 90% selectivity. The group used a combination of hydrogen peroxide and a ZSM-5 catalyst that had been doped with copper and iron. The development offers the opportunity to avoid the capital intensive production of synthesis gas as the first step in methanol production.

CHAPTER 9

Electric Vehicles

9.1 HISTORY

Electric vehicles are not new and are even older than those powered by diesel fuel that was introduced toward the end of the nineteenth century. Electric vehicles were developed in the mid-eighteenth century at a time when liquid fuels were rare. Electric vehicles differ from those powered by liquid fuels, whether fossil or renewable, since the energy they consume can be generated from a large variety of sources including solar and nuclear power. Although internal combustion engines for automobiles have been the preeminent source of power for such vehicles, electric power has long been commonplace for such types of transportation as railroad locomotives.

The use of electric power for transportation dates back to 1838 when Robert Davidson, a Scotsman, built a locomotive that attained a speed of 4 miles per hour. At about the same time, another Scotsman, Robert Anderson, introduced an electrically powered carriage that was powered by nonrechargeable primary cells. In a sense, his was the first electrically powered automobile. His efforts were followed in 1859 by those of Gaston Plante, a French physicist. Plante introduced the rechargeable lead-acid storage battery that is essentially still used today. Plante's efforts were improved upon by another Frenchman, Camille Faurer, who provided the capability for a battery to supply current and further developed it for use in automobiles. More than 30 years later, William Morrison of Des Moines, IA, built the first successful electric automobile in the United States.

In 1900, electric automobiles were the dominant vehicles and held many of the long distance and speed records. They were produced by unfamiliar companies such as Baker Electric, Detroit Electric, and Columbia Electric. They outsold gasoline powered automobiles, and in 1900, accounted for 28% of the 4192 cars that were in use in the United States. They were so popular that they constituted one-third of the automobiles in New York City, Boston, and Chicago. As late as

1913, the then President Woodrow Wilson and his Secret Service Agents toured Washington, DC, in their Milburn Electrics that could travel up to 70 miles without refueling.

The first electric taxis were on the streets of New York City in 1897, and the Connecticut-based Pope Manufacturing Company became the first large-scale producer of American automobiles. In 1900, the Lohner-Porsche Elektromobile debuted at the Paris Exposition. Although a purely electric vehicle Ferdinand Porsche soon added a combustion engine to recharge the batteries, a few years later, the Woods Motor Company introduced the Woods Dual Power, a four-cylinder gasoline-battery combination that had a top speed of 36 miles per hour. It was not a success. These efforts were the forerunners of what might be considered the hybrids of today. At nearly the same time, in his belief that electricity would become the dominant means of powering automobiles, Thomas Edison began an effort to create a powerful long-lasting battery for them. Although his research led to some improvements in the alkaline battery, he abandoned his efforts 10 years later. As noted later in this chapter, he was probably 100 years ahead of his time.

A number of developments contributed to the decline of the early electric automobile, notably improved highways that offered a greater driving range that could be attained by gasoline powered cars, and most importantly, the growth in the availability of low price gasoline that made such vehicles inexpensive to operate over long distances. In addition, the introduction of the electric starter eliminated the need for a hand crank to start a gasoline engine. Also, the development of the muffler lowered the noise generated by gasoline engines and made the use of such vehicles more bearable. Finally, there was the introduction of mass produced gasoline powered vehicles by Henry Ford. This significantly reduced the cost of gasoline powered automobiles compared to electric vehicles. Another major factor that contributed to the decline of electric vehicles was the gradual replacement beginning in the 1930s of electric powered trolley cars by liquid fueled buses in major cities. This is a phenomenon that is currently undergoing reversal for environmental reasons. In 1966, Congress introduced the earliest legislation to promote electric vehicles as a means to reduce air pollution. A Gallup poll at that time found that 33 million Americans were interested in electric vehicles. This interest was enhanced by the then soaring price of gasoline as a consequence of the OPEC and Iranian oil embargoes of 1973 and 1978.

9.2 BATTERIES

In batteries, electrical energy is produced by chemical reactions that result in products of lower energy content. The well-known lead-acid battery is an example of this and is commonly used in conventional gasoline powered vehicles in order to start the engine. The common lead-acid battery is an array of six 2-V cells that are connected in series to produce a 12-V battery. A compatible pair of electrodes or plates are joined through a sulfuric acid electrolyte and are further joined through an external wire through which electrons can pass from one plate to the other. The circuit is completed by the ions transferring across the cell via the electrolyte and the electrons moving through the external wire from the anode to the cathode. The reactions that take place when the cell is discharging and then again when it is charging are as seen in Figure 9.1.

In the discharged state, both positive and negative electrodes become lead sulfate and the electrolyte loses much of its sulfuric acid and becomes water primarily. In the charged state, a cell contains negative electrodes of elemental lead and positive electrodes composed of lead oxide, all contained in a sulfuric acid electrolyte.

Current electric vehicles are powered either by batteries as was done a century ago or by fuel cells. The major distinction is that in such vehicles that are battery powered the batteries are not the classical lead-acid type but are lithium batteries. These are environmentally

Negative plate reaction:

$$Pb(s) + HSO_4^-(aq) \rightarrow PbSO_4(s) + H^+(aq) + 2e^-$$

Positive plate reaction:

$$PbO_2(s) + HSO_4^-(aq) + 3H^+(aq) + 2e^- \rightarrow PbSO_4(s) + 2H_2O(l)$$

The total reaction can be written as

$$Pb(s) + PbO_2(s) + 2H_2SO_4(aq) \rightarrow 2PbSO_4(s) + 2H_2O(l)$$

Figure 9.1 Lead-acid battery half reactions.

friendly and have much higher energy density, considerably longer life and higher power than lead-acid batteries. The power in a vehicle driven by an electric motor is measured in kilowatts (KW) where 100 KW is about equal to 134 horsepower. Since most electric engines deliver full torque over a wide range of revolutions per minute (rpm), their performance is not equivalent but far greater than a 134 horsepower engine that has limited rpms.

The three primary components of a lithium battery are its positive and negative electrodes and an electrolyte, all of which are conceptually the same as in a lead-acid battery but with some major differences. Generally, the negative electrode is constructed of carbon, the positive electrode is a metal oxide, and the electrolyte is a lithium salt contained in an organic solvent. The electrochemical roles of the electrodes are reversible depending on the direction of the current flow through the cell. The negative electrode is usually graphite and the positive electrode is constructed of a mixed metal oxide such as lithium cobalt oxide. Organic nonaqueous materials, such as ethylene carbonate and diethyl carbonate, usually serve as electrolytes. These typically contain lithium perchlorate or lithium tetrafluoroborate. Depending on the materials of construction of the cell, its energy density, voltage, life, and the safe operation of the battery can differ markedly. Lithium, a Group I metal, is highly reactive when in contact with water and could be the cause of a fire.

Each of the two electrodes allows lithium ions to move into and out of their interiors. When a lithium cell is discharging, the positive lithium ion moves from the negative graphite electrode and enters the positive electrode. As in lead-acid batteries, the reverse occurs when the cell is charging as illustrated in Figure 9.2.

A typical 1 kg lithium battery can store 150 watt-hours of energy compared to 25 watt-hours for a similar size lead-acid battery. It therefore requires a 6 kg lead-acid battery to exhibit similar performance.

Positive electrode half reaction

$$LiCoO_2 \leftrightarrows Li_{1-x}CoO_2 + xLi^+ + xe^-$$

Negative electrode half reaction

$$xLi^+ + xe^- + xC_6 \leftrightarrows xLiC_6$$

Figure 9.2 Lithium ion battery half reactions.

Moreover, batteries with lithium−sulfur electrodes that are under development at the Chinese Academy of Sciences [81] are reportedly able to store as much as 4 times as much energy as conventional lithium ion batteries.

Battery powered vehicles are either plug-in or hybrids which use a combustion engine combined with battery power. Hybrid electric vehicles are powered by an internal combustion engine coupled with an electric motor in which energy is stored in batteries [79]. The extra power provided by the electric motor allows for a smaller engine. In addition, the battery can power auxiliary loads like headlights and sound systems. The battery can also reduce engine idling when the vehicle is stopped. Together these features result in better fuel economy without sacrificing performance. A hybrid vehicle is not able to plug into outside sources of electricity to recharge the battery. Instead, the vehicle uses regenerative braking and the internal combustion engine for recharging. The automobile uses energy normally lost during braking by using the electric motor as a generator and storing the captured energy in the battery. The engine recovers energy from the battery to provide extra power during acceleration.

Hybrids can be either mild or full hybrids and can be designed in either a series or parallel configuration. Mild hybrids are sometimes referred to as microhybrids and use a battery and an electric motor to help power the vehicle, and they can allow the engine to shut off when the vehicle is stopped at traffic lights and in stop-and-go traffic. This action further improves fuel economy. Mild hybrids are unable to power the vehicle by using electricity alone. Although these vehicles generally cost less than full hybrids, they also provide substantially poorer full economy benefits. Full hybrids, on the contrary, have more powerful electric motors and larger batteries and can drive the vehicle on only electric power although for short distances and at lower speeds. These vehicles cost more than mild hybrids but offer better fuel economy.

There are different ways to combine the power from the electric motor and the gasoline engine. Parallel hybrids are the most common design and connect the engine and the electric motor to the wheels through mechanical coupling. The electric motor and the gasoline engine each drive the wheels directly. In a series configuration, only the electric motor drives the wheels, and this configuration is sometimes found in plug-in hybrid vehicles.

Plug-in hybrid vehicles use batteries to power an electric motor and use another energy source such as gasoline or diesel to run the internal combustion engine. The use of electricity from a power grid to run the vehicle some or all of the time reduces operating costs and fuel consumption compared to conventional vehicles. Plug-ins also produce lower levels of emissions. Standalone plug-ins have larger battery packs than their hybrid counterparts. Accordingly, such vehicles can be driven from 10 to more than 40 miles on batteries alone. This makes it possible to use them for moderate driving without relying on an alternate fuel supply in cities. An increasing number of cities are providing a growing number of dedicated charging stations at public locations such as shopping centers. Since recharging usually requires several hours, it can be done while the vehicle owners are at work. In many US cities, the recharging stations are provided without cost to the owner of the vehicle.

The first full-sized, fully powered hybrid vehicle was introduced by General Motors in 1972. It was a Buick Skylark and was submitted by them as part of the 1970 Federal Clean Air Incentive Program. In a misguided action, Congress terminated this program in 1976. Vanguard-Sebring introduced its Citicar at the Electric Vehicle Symposium that was held in Washington, DC, shortly afterward. The little car was a runabout for city use as its name implied. It had a top speed of only 30 miles per hour and a range of only 40 miles. The company shut down in the late 1970s.

In 1975, the US Postal Service purchased 350 electric jeeps to be used in a test program for local deliveries. A year later, Congress passed the Electric and Hybrid Vehicle Research, Development and Demonstration Act. The intention of the legislation was to encourage the development of new technologies for the improvement of batteries, engines, and hybrid electric components.

A decade later in 1988, General Motors agreed to team up with California's AeroVironment to design what became the EV1. This was hailed as the "world's most efficient production vehicle." At about the same time, in 1990, Congress passed the zero emission vehicle (ZEV) mandate that required 2% of a state's vehicles to have no emissions by 1998, and to have no more than 10% by 2003. Like previous legislation that was terminated or weakened, this too was repeatedly weakened over the next decade.

Table 9.1 All-Electric Automobiles Produced in 1997–2000	
Honda EV Plus	GM EV1
Ford Ranger Pick-UP EV	Nissan Alta EV
Chevrolet S-10 EV	Toyota RAV4

Led by Toyota who introduced the Prius hybrid in 1997, a number of companies introduced a few thousand all-electric cars in the following years. Toyota sold nearly 18,000 cars in its first year of production. The all-electric cars produced by the others were mostly for lease, not for sale. This flurry of excitement was short-lived and their production ceased by 2000 (Table 9.1). Automakers clearly had little enthusiasm for all-electric cars. Toyota, to their credit, continued to forge ahead, and by mid-2013 had sold 3 million Prius automobiles worldwide. It continues to be the best selling hybrid [78]. Concurrently, Nissan has sold 100,000 of its Leaf model which has become the best-selling, highway capable, all-electric car in history [76,77].

GM and Daimler Chrysler sued the California Air Resources Board to repeal the ZEV mandate, and a year later, in 2003, they announced that they would not renew leases on their EV1 on the grounds that they were unable to supply parts for its repair. The following year, GM announced its intention to destroy the EV1s. This led to a public vigil to pressure GM to keep from demolishing about 78 EV1s. The effort failed. The following year, Tesla Motors unveiled its super sporty Roadster at the San Francisco International Auto show. The battery powered vehicle had a list price of $98,950.

Outside the United States in 2008, the Israeli government announced its support for battery powered vehicles in Israel. The incentive for their venture was the small size of the country and its lack of domestic fossil fuels. The strategy was to populate the country with *switching* stations where a charge-depleted vehicle could be quickly exchanged for a fully charged one in no longer time than would have been required to refill the gas tank of a conventional automobile. In view of the small size of the country, few such battery exchanges would be required in the course of the trip. The sponsoring company, Better Place, contemplated an analogous venture in Denmark. Subscribers to the program paid $350 per month to lease access to the battery service. Unfortunately, the concept failed because there were only 1000 battery powered vehicles and about a few hundred in Denmark, and consumers expressed little interest. Better Place filed for bankruptcy in mid-2013.

The US government tried again in 2008–2009 to promote electric vehicles by enacting legislation wherein buyers of battery powered cars after December 31, 2009, were entitled to a tax credit of $2500 per vehicle. Cars had to be driven by a battery that had at least 5 kilowatt hours (kWh) of capacity. Purchasers were entitled to an additional $417 for each kWh of battery capacity in excess of the base 5 kWh. The total amount of credit per vehicle was a maximum of $7500. The credit was phased out once a manufacturer had sold at least 200,000 of a qualifying vehicle in the United States. The credit was for new purchases only and did not extend to resales.

In terms of fuel costs based on electricity rates in 50 US cities, a study [80] completed by The Union of Concerned Scientists concludes that drivers of electric vehicles will spend $750–1200 less per year than those who operate an average new compact car fueled by gasoline at $3.50 per gallon. Despite on and off again legislation in the United States, there is perceptible growth in electric vehicles, their small market size notwithstanding. Apparently, we are in the early part of the classical S-shaped growth curve for new products. The US Department of Energy projected that about 45,000 battery powered vehicles were produced in 2011. In addition, the global market for lithium ion batteries is projected to grow from $5 billion to $47 billion by 2020 according to ABI Research, a technology market research firm.

9.3 FUEL CELLS

No technology ever exists in a vacuum, and battery powered automobiles, hybrid or not, are no exception. Fuel cells offer a competitive alternative. Despite its name, a fuel cell does rely on a continuous supply of fuel, which most commonly is hydrogen. The hydrogen reacts with either oxygen or air to produce electric power continuously as long as they are supplied. Like most innovations, fuel cells have a long history and were introduced nearly 200 years ago in 1838. Like batteries, fuel cells have a cathode, an anode, and an electrolyte that enables a charge to move between the two sides of the cell. The principal difference between the various types of fuel cells is the type of electrolyte followed by the difference in start-up time. After languishing for more than 100 years, fuel cell development began in earnest in the 1960s with a collaborative effort between NASA and the General Electric Company. The first fuel cell powered automobiles weren't developed until 1991.

In a hydrogen powered fuel cell, hydrogen as a gas is oxidized at the anode to form hydrogen ions and electrons. The electrolyte is specifically designed to enable hydrogen ions to pass through it to the cathode but not the electrons which travel through an external wire to the cathode. At the cathode, the hydrogen ions react with the oxygen in the cell to produce water. The anode is usually a finely divided platinum powder. The cathode, in turn, is commonly composed of nickel. A schematic diagram of a proton conducting fuel cell is shown in Figure 9.3.

The US Department of Energy [83,84] estimates that an 80 kW automotive fuel cell might be produced for $67 per kW on the basis that 100,000 such units would be produced. A lower cost of $55 per kW was projected if as many as 500,000 units per year were produced. They used cells that had proton exchanging membranes constructed of Nafion (perfluorosulfonic acid) and had a 40−60% energy efficiency and an

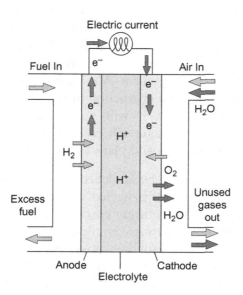

Figure 9.3 Hydrogen fuel cell. R. Dervisoglu in Wikimedia Commons; http://en.wikipedia.org/wiki/File:Solid_oxide_fuel_cell.svg.

operating temperature of 50−100°C. The energy efficiency is measured by the ratio of useful energy put into the system compared to the amount of useful energy produced by it. This is higher than the efficiency of an internal combustion engine whose efficiency is usually about 25%. The maximum theoretical energy efficiency of a fuel cell is 83% when operating at a low power density and with pure hydrogen and oxygen as reactants. In practice, the tank-to-wheels efficiency of a fuel cell vehicle is greater than 45% at low loads. Fuel cells cannot store energy like a battery except for its contained hydrogen. If one takes into consideration losses from steam methane reforming to produce the hydrogen, its storage and shipping into account, efficiency can be as low as 22%.

Several automobile producers [85] are introducing fuel cells into their 2015 model year cars, notably Nissan, Honda, Toyota, and Hyundai. This is a large change for fuel cells that had been shunted aside in the last decade in spite of the fact that the vehicles they powered were nonpolluting. Such cars emit only water vapor from their tailpipes. Producers of fuel cell powered vehicles had been focusing on plug-in electric cars. A factor contributing to their sudden attractiveness besides their nonpolluting character is the advent of large volume, low priced natural gas as a source of hydrogen. When Nissan introduced its new SUV at the 2013 Paris Motor Show, it claimed that fuel costs would be one-sixth those of their previous estimates. In preparation for the growth of fuel cell powered vehicles, Germany [86] is planning to install a network of about 400 hydrogen refueling stations across the nation. The initiative is supported by a number of major hydrogen producers, among them Air Liquide, Linde, Shell, Total, and OMV. The cost for the construction of the hydrogen infrastructure is about $475 million, and it will not be completed until 2023. Similarly, the State of California will invest $200 million for 20 hydrogen refueling stations in 2015 and plans to have 100 of them by 2024.

The 2015 fuel cell cars to be launched by Toyota culminated a 20-year quest during which they struggled to make the technology work and strived to lower manufacturing costs to permit realistic pricing in the $30,000−40,000 range after government subsidies. This places them in direct competition with Honda whose FCX Clarity test fleet has been on the streets for 6 years. Both cars will be competing with battery powered vehicles such as Nissan's Leaf and Tesla. In order to ensure its competitiveness with other lithium ion battery powered vehicles, Tesla formed a partnership with Simbol (Pleasanton, CA) for a lithium ion battery plant scheduled to open in 2017. Adequate supply of lithium is essential for all battery powered cars whose dependence on lithium accounts for 35% of lithium production. Lithium consumption has been increasing at an annual rate of 6.4% since 2000 according to the US Geological Survey. Simbol [87] has developed novel technology for its recovery from wastes generated by geothermal power plants. Toyota, in the meantime, had invested $50 million in Tesla in 2010 for a 3% share of the company. In addition, Toyota had also signed a $100 million joint development agreement with Tesla for a joint effort involving Toyota's RAV4 SUV that would employ Tesla's electric power train. In May 2014, Toyota appeared disenchanted with battery powered vehicles and reportedly was prepared to walk away from the Tesla agreements in favor of fuel cell technology. Toyota appears to regard fuel cells as superior to battery powered systems [89].

Inexpensive hydrogen is a prerequisite for the success of hydrogen powered fuel cells. Current technology is essentially reliant on steam methane reforming to produce hydrogen and this returns us to fossil, not renewable, fuels as starting materials. Electrolysis of water continues to be an uneconomical way to produce hydrogen. In recognition of this, for 5 years, the Department of Energy has been funding work at Arizona State University in the development of a *fake leaf* as part of their bio-inspired solar fuel research effort. The intent is to produce hydrogen by artificial photosynthesis using solar energy [102]. Funding has been $14.5 million as of 2014.

The US Department of Energy has launched an online tool to promote electrical vehicles by comparing their fuel costs against those that are gasoline powered. The online tool offers a state-by-state comparison of the two vehicle types by introducing an eGallon [88]. An eGallon in Pennsylvania is $1.33 and means that a typical electric vehicle could travel as far on $1.33 of electricity as a similar vehicle could go on a gallon of gasoline. In New Jersey where gasoline is more expensive, an eGallon was $1.55. In 2013, the average cost nationwide for an eGallon was a $1.17 in that year when the US average for gasoline was $3.68 per gallon.

CHAPTER 10

The Uncertain Future

In the last 50 years, energy research and development have undergone major changes in direction all of which may reoccur. Such changes were driven by political events that disrupted the economy at least twice in the last 40 years and may do so yet again. Major factors were real or threatened disruption in the supply of oil, the belated recognition of the threat of global warming, and finally, the health hazards associated with fuels in common use. Typical examples were a series of legislative acts that governed the importation of oil, the rush to develop coal as a petroleum alternative, and the introduction and then removal of octane improvers for gasoline.

10.1 COAL

In 1974 and again in 1977, the United States experienced two economy shaking oil embargoes, the first by OPEC and the second by the country of Iran. These occurred during a time when the United States was heavily reliant on oil imports that had peaked at 8565 barrels per day in 1977 and the country was striving to establish security against any such further events. The domestic focus turned to coal as an obvious solution since the United States, then as now, had huge coal reserves and struggled to determine how these could be used advantageously. Efforts to develop alternate fuels in the 1970s and 1980s were similar to those based on the use of cornstarch and sugarcane. There was ample history to draw on since coal had been the raw material of last resort by Germany during World War II and subsequently by the Union of South Africa. In each instance, these countries had sizable reserves of coal and little to no access to crude oil. In each instance, coal gasification coupled with Fischer Tropsch chemistry was the technology of choice. To a limited extent, the United States also considered coal gasification to produce a low BTU pipeline gas for power plants. This gas was essentially the same as the so-called *town gas* of the 1890s. Most efforts, however, were directed toward the production of liquid fuels for motor vehicles. This was at a time when the issue of energy security was paramount and there was no consideration of environmental issues such as global warming. There was no thought given to the development of renewable fuels except in Brazil which had substantial resources in the production of sugarcane.

The US government and the public endorsed coal development along with implied or actual financial support on the state and federal level. An aggressive effort was initiated to upgrade coal to liquid fuels. Exxon Corp. (now ExxonMobil), the largest domestic and global oil company, developed their *Donor-Solvent Process* in which ground-up coal was reacted at 450°C under 2000 psi pressure with hydrogen derived from a contained solvent to produce a gas and an oil. In 1974, Exxon commissioned a 200 ton per day pilot plant that was followed by a commercial facility that operated from mid-1980 until it was shut down in 1982. During this time, the plant successfully processed 90,000 tons of coal. With the termination of the Iranian oil embargo, the technology was abandoned.

Concurrent with Exxon's efforts, the Hydrocarbon Research Corporation of Trenton, NJ, developed their H-Coal process that was an elaboration of their H-Oil technology for the distillation of mineral oil residues. In this system, ground-up and dried coal was also mixed with a solvent that was derived from the process, and the resulting slurry was heated at 455°C under 3000 psi hydrogen pressure. The process used an ebullated bed reactor that produced naphtha, coal oil, and a solid containing residue. Like the Exxon effort, a 600 ton per day plant was constructed that operated from May 1980 to November 1982. It was then shut down and this technology too was shelved.

Not to be outdone, Gulf Oil Company (now part of Chevron) introduced their Solvent-Refined Coal (SRC) process that had similarities to the Exxon and Hydrocarbon Research efforts. Gulf Oil's effort was used in a 30 ton per day pilot plant that was operated until 1981, and like the other systems, was then shut down and the technology was also eventually abandoned.

Of considerable interest during the time that coal hydrogenation was being pursued was the development of coal-water slurries. As the name suggests, these were slurries of finely divided particles of coal (10−65 μm in size) suspended in water that had the appearance and burning characteristics of a no. 2 fuel oil. Generally, the slurries were composed of 50−70% lignite and the balance being water with occasional addition of additives to stabilize the suspension. They were originally developed by the Soviet Union at Novosibirsk and explored by the United States and others worldwide. The slurries could be readily shipped by pipeline like any oil and were attractive as fuel for power plants. Their attractive feature was that they could be easily pumped through existing pipelines and injected into furnaces and boilers. Ceramic nozzles were used to accommodate the abrasive nature of the contained coal. Slurries whose coal particle size was less than 20 μm were also found to be suitable as diesel fuel replacements.

Interest in coal-water slurries continues in countries like China and India where coal is a major factor for power plant operations. Although they constitute a convenient way to handle coal, they are nonetheless still coal with the same negative environmental characteristics. If not for the threat of global warming, they might be more broadly used currently. Coal, which was once king as an energy source, is now in the process of being gradually dethroned.

10.2 GASOLINE

Less than 100 years ago, gasoline was an unwanted byproduct of the infant petroleum industry. In the early twentieth century, it was used largely in the production of lubricating oils, greases, and stove and lamp oils. As late as 1910, gasoline powered vehicles were in a minority compared to electric cars and were regarded as a pastime for the wealthy. Today, gasoline engines for American vehicles produce more than 50 times the energy of all the nation's power plants. In 2013, US gasoline consumption was 135 billion gallons. This corresponds to a daily average of 369 million gallons and is only 6% less than the record high of 142 billion gallons in 2007. There are now a number of reasons to believe that gasoline may have reached its high water mark and may be in retreat.

To satisfy the public's demand for high-performance engines, gasoline must meet exacting specifications such as engine knock resistance as monitored by its octane number. This is essential to prevent annoying, fuel wasting, and damaging engine knock. A high octane number is therefore desirable throughout the distillation range of the gasoline. Ideally, the fuel−air mixture in each cylinder of the engine should burn smoothly and evenly following the timed ignition by the spark plug. At times the spreading flame front sweeps across the combustion chamber unevenly and the unburned portion is so heated and compressed that it auto-ignites and detonates in an instant. As a result, instead of pushing downward in a power stroke, a piston gives a hard instantaneous rap to which it can't respond because it is connected to other pistons that are going through different phases of the prevailing engine cycle. This abrupt release of energy causes high-frequency fluctuations throughout the combustion chamber that sounds like a sharp metallic noise called a knock. Fuel energy that should have been converted to useful power is instead dissipated in the form of a pressure wave and increased heat to the surrounding engine parts. Besides an objectionable banging sound and a waste of energy, prolonged knocking overheats the engine. This potentially shortens its life and damages the engine. In essence, the presence or absence of a knock depends on the race between the advancing flame front and precombustion reactions in the heated gas [90,91].

It was found that engine knock could be suppressed by the addition of tetraethyl lead to the gasoline supply. The addition of such lead compounds was universally adopted by all gasoline producers worldwide. It subsequently became apparent that some measuring technique was essential for rating the antiknock quality and power of a gasoline. To achieve this, n-heptane, a linear paraffin that burns with a considerable degree of knock, was chosen as the bottom of the scale and designated with an octane number of zero. At the top of the scale, a highly branched nonknocking hydrocarbon, 2,2,4-trimethylpentane, was chosen and assigned an octane number of 100. The octane number of a gasoline was thereupon one which matched a particular blend of 2,2,4-trimethylpentane and n-heptane. If a gasoline corresponds to a blend of 67% of the former and 33% of the latter, it is described as having a 67 octane number.

After 50 years of use, tetraethyl lead was phased out in the United States in the 1970s because of a growing awareness of its cumulative neurotoxicity plus the damaging effect it had on the catalytic mufflers whose purpose it is to control pollution. Tetraethyl lead is discharged from the tailpipes of vehicles and accumulates in the body at low levels of exposure and is especially harmful to children. After its termination in the United States, the average blood level in Americans dropped

72% from 16 to only 3 μm per deciliter. Tetraethyl lead is no longer used in most countries with the exception of Myanmar, North Korea, and several in the Middle East.

With the termination of tetraethyl lead, gasoline producers quickly moved to methyl-*t*-butyl ether (MTBE) as a substitute. In addition to functioning as an octane improver, MTBE helps gasoline burn more completely and thereby reduces tailpipe emissions. The presence of oxygen in the MTBE molecule optimized engine combustion. For a while it was regarded as a panacea of sorts for gasoline. Unfortunately, the use of MTBE was relatively short-lived because, like tetraethyl lead, it too, was found to have environmental and health hazards. MTBE was belatedly found to pollute groundwater from gasoline spillage and from leaks in underground storage tanks. An early indication of its likely demise were reports by drivers in Alaska who became dizzy when filling their gas tanks during the winter months. This alarmed many people [93] and, among other things, initiated calls for its use to be terminated. Cost estimates [92] for its removal from groundwater were as great as $30 billion. California and New York State led in MTBE consumption by accounting for 40% of the US total in 2004 and the two states banned its use in gasoline. They were followed by an additional 25 states in 2005. MTBE had a life span of only 30 years, and at its highpoint in 2002, it had a global demand of nearly 22 million tons. It held center stage in the oil and chemical industries for nearly two decades, as companies raced to introduce it into their gasoline. A number of processes were introduced for its manufacture as were a number of plants. Shortly after the termination of MTBE, Congress enacted the Energy Policy Act (Chapter 1) that removed the oxygenate requirement for reformulated gasoline. At the same time, Congress also instituted the Renewable Fuels Standard (RFS) following which refiners quickly began to blend ethanol (Chapter 2) into gasoline. MTBE has not been used in any appreciable quantity since 2005.

10.3 THE PRESENT AND THE FUTURE

Despite the efforts of the United States and other countries to limit its use, coal continues to be the fastest growing energy source worldwide. It continues to grow at a rate of 2.3% per year, and according to the Paris-based International Energy Agency, it will continue to do so through 2018 when it may dethrone crude oil [94] as the major source of supply. Such growth in demand is driven by China which is the world's largest producer and consumer of coal. China accounts for 46% of global coal production and 49% of consumption—nearly as much as the rest of the world combined. Led by China, the 10 largest coal producing countries supply 90% of the world's coal. China itself produces nearly four times as much coal as the second largest producer, the United States. who accounts for 12% of global production. Coal accounts for most of China's energy consumption and has maintained a 70% share of it since 1980. China's domestic natural gas production has been described as "stretched to the limit," but the country has vast natural gas reserves including shale [95]. China's reliance on coal may change dramatically in the future as a consequence of their execution of a 30-year agreement with Russia in May 2014. The agreement notwithstanding, actual deliveries may be constrained for some time because there is no preexisting pipeline to China from the Russian gas fields.

Coal accounts for only 18% of US energy consumption. The US Energy Information Administration expects that total US consumption will grow only about 8% in the next 28 years, mostly for electric power generation. Little new coal-based capacity is expected to be added as power plants gradually shift to natural gas. Many coal-fired power plants (40 gigawatts of capacity) are expected to be retired by 2020. These plants account for about one-sixth of US coal capacity and about 5% of total electricity generation nationwide. Such plants are sensitive to natural gas prices. Natural gas is the marginal fuel for power generation.

Ongoing developments in energy are expected to reshape the US picture for vehicle fuels. With an annual growth rate of 0.8 million barrels per day through 2016, crude oil production in the United States is expected to reach a volume of 9.5 million barrels per day and then level off and slowly decline through 2020 and thereafter. Natural gas production is projected [97] to increase steadily with a 56% increase to 2040 when production will be 37.6 TCF. Such growth in natural gas production is likely to have its greatest impact on renewable fuels development for which it is both complimentary and competitive. Many renewable fuels processes that employ synthesis gas can either continue to obtain the latter by gasification of biomass or abandon it in favor of natural gas in some combination of a biomass or synthetic fuels system. Natural gas is expected to overtake coal by 2035 to account for the major share of US electric power generation. In 2035, natural gas is expected to account for 35% of such power generation capacity while coal declines to 32%. Generation from renewable fuels, unlike coal and nuclear power, will continue to increase. Renewable fuels are likely to continue to benefit from legislation enacted at the beginning of 2013 when tax credits for them were instituted.

The benefits to be gained from biofuels produced from waste cellulose came under a cloud when a Federally funded study surprisingly concluded that such fuels caused more greenhouse gas pollution than unreformulated gasoline [97]. The $500,000 study that was published in the journal, *Nature Climate Change*, concluded that biofuels produced from corn residue release 7% more greenhouse gases in the early years. The conclusion strikes at the heart of the ongoing effort to produce cellulosic derived ethanol and suggests that such biofuels should be ineligible as a candidate under the Reformulated Fuels Standard. The study, which was carried out at the University of Nebraska-Lincoln, concludes that the use of corn stover makes undesirable changes to the carbon composition of the soil. As would be expected, the study has been criticized by parties in industry and in academia that fault a number of its premises.

In addition to increases in domestic consumption, US exports of natural gas are expected to increase. Exports of LNG will probably rise to 3.5 BCFD per year by 2029 and remain at that level through 2040. Pipeline exports will probably increase 6% annually from 0.6 to 3.1 TCF between 2012 and 2040. Exports to Canada are forecast to increase 1.2% per year but pipeline imports from Canada may decline 30% as US reserves increase. Coincidentally, imported fuels of all kinds will also fall with domestic increases of oil and natural gas. Imports as a percentage of US consumption fell from 30% in 2005 to 16% in 2012 and will probably be only 4% in 2040.

For the first time since 1990, carbon dioxide emissions associated with industrial activity including power plant generation are expected to surpass those from transportation by the middle of the next decade. This is the result of new vehicle economy standards, biofuel mandates, and changes in consumer behavior.

Diesel fuel is projected to surpass gasoline as the primary US transportation fuel by 2020 in the view of ExxonMobil [98] and other oil companies such as Valero Energy. Diesel will continue to increase its market share for vehicles at least through 2040. The shift away from motor gasoline to diesel is driven by improvements in light-duty vehicles and growth in commercial transportation. Diesel demand is expected to account for 70% of the growth in all types of transportation. Fuel demand for heavy-duty vehicles, the largest subsector, will see the greatest growth by increasing 65% and may account for all transportation fuels by 2040. About 80% of the growth in diesel is expected to come from developing countries. Diesel demands for aviation and marine uses are expected to increase their combined share from 20% to 25% by 2040. ExxonMobil's view is also shared by J.D. Power & Associates and by the Freedonia Group [99], a Cleveland, Ohio market research firm, both of whom project an increase of 7.7% per year in diesel demand by 2017. They conclude that the increased production of heavy trucks and buses and off-highway equipment will combine with a rebound in light vehicle production in western Europe will be important factors. Underlying the growth in diesel demand is the growing popularity of such cars in the United States and India. Their collective views are summarized in Table 10.1.

The Asia Pacific Region is and will continue to be the largest producer of diesel vehicles, more than double that of Western Europe and greater than North America in 2017 by a factor of nearly four. This is attributable to the Japanese, South Korean, Chinese, and Indian markets that continue to be the fastest growing group of diesel automobiles worldwide. Western European manufacturers of diesel vehicles are

Table 10.1 Global Diesel Engine Demand[a] (billion dollars)

	2007	2012	2017
North America	18.6	23.2	32.3
Western Europe	54.8	43.3	57.2
Asia Pacific	43.6	74.3	113.5
South America	7.1	9.4	14.2
Eastern Europe	11.5	13.2	18.7
Africa/Middle East	6.3	8.2	12.7
Total	141.9	171.6	248.6

[a]World Diesel Engines; Freedonia Group.

recognized as the gold standard and the growing popularity of their automobiles combined with the rising incomes in developing countries will further boost their growth in production. Several US producers, such as Chrysler's Grand Cherokee, GM's Cadillac ATS, and the Chevrolet Cruze have first-time offerings of diesel models in 2014.

The United States imports two types of biomass-based diesel, fuel biodiesel, and renewable diesel. The latter is a diesel-like fuel that is compatible with the infrastructure for existing engines in any blending proportion. It is produced by hydrotreating vegetable oils (mainly soybean) and animal fats. Biodiesel can be produced by the processes described in Chapters 5, 6, and 7. The US diesel imports increased nearly nine-fold [100] in the 2012–2013 period from 61 million to 525 million gallons. The rapid growth in US biomass-based diesel demand is a result of the increasing mandate for it under the RFS. Both fuel biodiesel and renewable diesel qualify for the RFS targets described in Chapter 1. This stipulates an increase from 15.2 to 16.55 billion gallons from 2012 to 2013. Biomass-based diesel also qualifies for the California Low Carbon Standard. This sets annual carbon intensity targets for fuel providers to reduce the carbon content of gasoline and diesel fuels through 2020. Fuels with low carbon intensity values generate credits for fuel supplies that can offset deficits accumulated from fuels with higher values.

In contrast to diesel, global gasoline demand is not expected to increase very much at all after 2014 even though the number of personal motor vehicles is expected to double in size from 800 million to 1.6 billion by 2040. Although increasing in demand at the expense of gasoline, as more options such as electric vehicles and hybrids become available, gasoline and diesel together will experience a declining share of the vehicle market.

The United States became a net exporter of petroleum products in 2012, and as described above, diesel became the export product of choice because domestic profit margins are significantly greater than those of its foreign counterparts. As of 2014, gross margins for US Gulf Coast producers were nearly three times as great as those for gasoline, $15 versus $5 per barrel according to Valero Energy as shown in Figure 10.1. As a result, there has been a race to add diesel capacity at twice the rate of gasoline. Global diesel fuel demand in recent years is illustrated in Figure 10.2. Diesel fuel can be produced by a number of

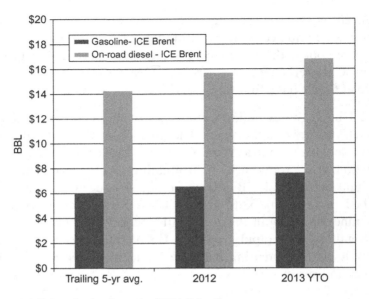

Figure 10.1 US Gulf Coast diesel profit margins (2013). Valero Energy.

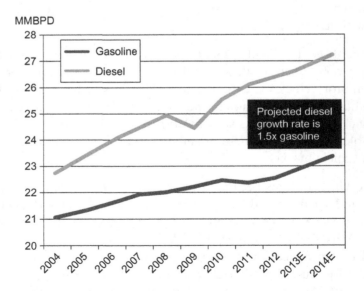

Figure 10.2 World product demand. Source: Consultant (EIA and IEA) and Valero estimates. Consultant annual estimates generally updated 6–12 months after year end.

processes, such as from algae, synthesis gas conversion and Fischer Tropsch synthesis.

By 2025 ExxonMobil [98] expects that hybrid vehicles will be less expensive and their market share will expand quickly, and thereby adding more efficiency to motor vehicles as a group. They project that full hybrid vehicles will account for 40% of the total in 2040 and will account for a full 50% of new car sales in that year. They expect the aggregate share of fossil fuels to decline from 82% to 78% by 2040. In contrast, they believe that renewable fuels will grow rapidly from 9% to 13% of total consumption. Again, a major factor in the outlook for renewable fuels demand is the continued implementation of the RFS. The continuing production of natural gas from shale may inhibit application of this legislation.

US Cornstarch Fermentation Ethanol; Production Capacity (2013)*

Company	Location	Million Gallons
Abengoa Bioenergy	York, NE; Colwich, KS;	198 (total)
	Portales, NM; Ravenna, NE	
ACE Ethanol	Stanley, WI	39
Adkins Energy	Lena, IL	40
Advanced Bioenergy	Fairmont, NE	100
AGP	Hastings, NE	52
Agra Resources	Albert Lea, MN	48
Agri Energy	Luverne, MN	21
Alchem Ltd	Grafton, ND	11
Al-Corn Clean Fuel	Claremont, MN	35
Amaizing Energy	Denison, IA	40
Archer Daniels Mid.	7 plants; (153 mill. gal avg)	1070
Aventine Renewable	Pekin, IL	157
	Aurora, NE	50
Badger State	Monroe, WI	48
Big River Res.	West Burlington, IA	40
Broin Enterprises	Scotland, SD	9
Bushmills Ethanol	Alwater, MN	40
Cargill	Blair, NE & Eddyville, IA	100
Central Indiana	Marion, IN	40
Central MN Ethanol	Little Falls, MN	22
Central Wisconsin	Plover, WI	4
Chief Ethanol	Hastings, NE	62
Chippewa Valley	Benson, MN	45
Commonwealth	Hopkinsville, KY	33
Corn LP	Goldfield, IA	50
Cornhusker Energy	Lexington, NE	40

(*Continued*)

*U.S. Energy Information Agency

(Continued)		
Company	**Location**	**Million Gallons**
Corn Plus	Winnebago, MN	44
Dakota Ethanol	Wentworth, SD	50
DENCO	Morris, MN	22
E3 Biofuels	Mead, NE	24
East Kansas Agri.	Garnett, KS	35
Ethanol2000	Bingham Lake, MN	32
Frontier Ethanol	Gowrie, IA	60
Front Range Energy	Windsor, CO	40
Glacial Lakes	Watertown, SD	50
Golden Grain Ener.	Mason City, IA	40
Golden Triangle	Craig, MO	20
Grain Processing	Muscatine, IA	20
Granite Falls Ener.	Granite Falls, MN	45
Great Plains Ethanol	Chancellor, SD	50
Green Plains Renew.	Shanandoah, IA	50
Hawkeye Renewable	Iowa Falls & Fairbanks, IA	150
Heartland Corn Pdts.	Winthrop, MN	36
Heartland Grain Fuel	Aberdeen & Huron, SD	39
Heron Lake Bioen.	Heron Lake, MN	50
Horizon Ethanol	Jewell, IA	50
Husker Ag	Plainview, NE	27
Illinois River	Rochelle, IL	50
Iowa Ethanol	Haniontown, IA	50
Iroquois Bioenergy	Rensselaer, IN	40
James Valley Eth.	Groton, SD	50
KAAPA Ethanol	Minden, NE	40
Lincolnland Agri.	Palestine, IL	48
Lincolnway Energy	Nevada, IA	50
Little Sioux Corn	Marcus, IA	52
MGP Ingredients	Pekin, IL	78
Michigan Ethanol	Caro, MI	50
Mid America Agri	Madrid, NE	44
Mid-Missouri Ener.	Malta Bend, MO	50
Michigan Grain	Lakota, IA & Riga, MI	152
Midwest Renewable	Sutherland, NE	22
Minnesota Ener.	Buffalo Lake, MN	18

(Continued)

(Continued)		
Company	Location	Million Gallons
Missouri Ethanol	Laddonia, MO	45
New Energy	South Bend, IN	102
North Country	Rosholt, SD	20
Northeast Missouri	Macon, MO	45
Northern Lights	Big Stone City, SD	50
Northstar Ethanol	Lake Crystal, MN	52
Otter Creek Ethanol	Ashton, IA	55
Panhandle Energies	Dumas, TX	30
Phoenix Biofuels	Goshen, CA	25
Pine Lake Corn	Steamboat Rock, IA	20
Platte Valley Fuel	Central City, NE	40
Prairie Ethanol	Loomis, SD	60
Prairie Horizon	Phippsburg, KS	40
Pro-Corn	Preston, MN	42
Quad-County Corn	Galva, IA	27
Red Trail Energy	Richardson, ND	50
Redfield Energy	Redfield, SD	50
Reeve Agri-Energy	Garden City, KS	12
Siouxland Energy	Sioux Center, IA	25
Siouxland Ethanol	Jackson, NE	50
Sioux River Ethanol	Hudson, SD	55
Sterling Ethanol	Sterling, CO	42
Tall Corn Ethanol	Coon Rapids, IA	49
Tate & Lyle	Loudon, TN	67
The Andersons Albion	Albion, MI	55
Trenton Agri Products	Trenton, NE	45
United WI Grain Prod.	Friesland, WI	49
US Bioenergy	Albert City, IA & Lake, MI	145
Utica Energy	Oshkosh, WI	48
Val-E Ethanol	Ord, NE	45
VeraSun Energy	Aurora, SD & Ft. Dodge, IA	230
Voyager Ethanol	Emmetsburg, IA	52
Western Plains	Campus, KS	45
Western Wisconsin	Boyceville, WI	40
Wyoming Ethanol	Torrington, WY	5
Xethanol Biofuels	Blairstown, IA	5

BIBLIOGRAPHY

[1] B.D. Yacobucci, Analysis of Renewable Identification Numbers in the Renewable Fuels Standard. Congressional Research Service, November 16, 2012.

[2] New York Times, September 15, 2013, p. 1.

[3] Food and Agricultural Policy Research Institute, University of Missouri Report No. 07-09 (September 2009); Renewable Identification Number Markets.

[4] D. Pimental, T.W. Patzek, Nat. Resour. Res. 14(1) (2005) 43.

[5] M. Shapouri, M. Wang, J.A. Duffield, The Energy Balance of Corn Ethanol: An Update. Agricultural Economics Report No. 813, July 2002.

[6] U.S. Information Administration, October 18, 2012.

[7] Cornell University News, August 6, 2001.

[8] New York Times, March 17, 2015, p. 16.

[9] U.S. Energy Information Administration, March 11, 2013.

[10] A. Pander, Global Automakers, May 16, 2013.

[11] U.S. Energy Information Administration, October 15, 2012.

[12] Chemical & Engineering News, August 27, 2012, p. 26.

[13] Encyclopedia of Chemical; Technology (Kirk Othmer), 9, 132–153.

[14] <http://en.wikipedia.org/wiki/Ethanol_fuel_in_Brazil>.

[15] C.T. Montgomery, M.B. Smith, Hydraulic fracturing: history of an enduring technology, J. Petrol. Technol. (2010).

[16] Today In Energy, U.S. Energy Information Agency; U.S. Dept. of Energy, June 10, 2013.

[17] Chemical & Engineering News, October 21, 2013, p. 27.

[18] U.S. Energy Information Administration, November 15, 2013.

[19] Today In Energy, U.S. Energy Information Admin., July 25, 2013.

[20] BP Energy Outlook 2030, 2013 Edition Article.

[21] Hydrocarbon Processing.Com Online Article, November 18, 2013.

[22] Chemical & Engineering News, August 19, 2013, p. 7.

[23] N.C. Mosier, et al., Bioresour. Technol. 96(6) 673.

[24] <WWW.Zeachem.Com/technology/overview.php>.

[25] U.S.Patent 5,730,837 (March 24, 1998) to Midwest Process Research.

[26] W. Lan, C.F. Liu, R.C. Sun, J. Agric. Food Chem. 59 (2011) 8691.

[27] K.T. Klasson, et al., Enzyme Microb. Technol., 14, p. 602.

[28] A.M. Henstra, et al., Curr. Opin. Biotechnol. 18 (2007) 200.

[29] U.S. patent 8,080,693 to Enerkem, Inc., December 31, 2009.

[30] U.S. patent applic. 2012/0181483 (July 19, 2012) to Sundrop Fuels.

[31] New York Times, August 18, 2013, p. 5.

[32] Op. cit. en.wikipedia.org/wiki/Ethanol-fuel-in-Brazil.

[33] The Economist, March 3, 2007, p. 44.

[34] World Patent Applic. 20133083816 to Shell Oil Co.

[35] T. Jeffries, Y.S. Jin, Appl. Microbiol. 63(5) (2004) 495.

[36] B. Jiang (Ph.D. thesis). Washington University, 2006.

[37] U.S. Patent Applic. 20140045227 (February 13, 2014) to Clariant Produkte.

[38] U.S. Patent Applic 20140017751 (February 12, 2014) to Clariant Produkte.

[39] NY Times, October 24, 2012.

[40] Hydrocarbon Processing, September 2012, p. 79.

[41] British Patent 191504845, 1816 to C. Weizmann.

[42] U.S. 8,222,017 to Butamax (July 17, 2012) and U.S. Patent 8,273,558 (September 15, 2012) to Butamax.

[43] U.S. Patent Application 20120208246 (August 16, 2012 (to Butamax).

[44] P. Peralta-Yahya, et al., Nature 488 (2012) 320.

[45] U.S. Patent 8,378,160 B2 (February 19, 2003) to Gevo.

[46] J.L. Avalos, et al., Nat. Biotechnol. (2013). Available from: http://dx.doi.org/doi:10.1038/nbt.2509.

[47] E. Gak, et al., J. Ind. Microbiol. Biochem. (2014).

[48] P.H. Pfromm, Bioresour. Technol. 102 (2011) 1185.

[49] Hydrocarbon Processing, December 2012, p. 11.

[50] <www.oilgae.com/algae/cult/pbr/pbr.html>.

[51] Western Farm Press, November 29, 2011.

[52] U.S. Patent 7,691,159 (December 19, 2007) to Applied Research Associates.

[53] M. Calvin, J.A. Bassham, The Photosynthesis of T.N. Carbon Compounds, W.A. Benjamin, Inc., 1962.

[54] P. Schlegermann, et al., J. Combustion (2012).

[55] K.M. McGinnis, et al., J, Appl. Phycol. 9 (1997) 12.

[56] Energy Fuels, 20 (2) (2006) p. 812.

[57] T.N. Kelnes, et al., in Biofuels Technology, p. 7.

[58] U.S. Patent 8,216,182 B2 (July 26, 2011) to Joule Unlimited.

[59] Today In Energy U.S. Energy Information Administration, April 14 (2014).

[60] Outlook for Energy; A View to 2040, Exxon Mobil Corp, March 3, 2013.

[61] U.S. Energy Information Administration, February 13, 2014.

[62] S. Haus, et al., BMC Syst. Biol. 5 (2011) 10.

[63] U.S. Patent 5,611,214 to Battelle Memorial Institute, July 29, 1994.

[64] Chemical & Engineering News, December 16, 2013, p. 13.

[65] U.S. Energy Information Administration; March 31, 2014.

[66] Chemical & Engineering News, December 9, 2013, p. 20.

[67] Hydrocarbon Processing, April 1, 2012.

[68] K. Cohan in ExxonMobil Perspectives, March 22, 2012.

[69] U.S. Energy Information Administration, Monthly Energy Review, June 2011.

[70] U.S. Energy Information Administration, April 22, 2014.

[71] Navigant Research Report, Market Data: Natural Gas Vehicles, Second Quarter 2013 and July 18, 2012.

[72] U.S.A. Today, February 18, 2014, p. 1B.

[73] American Clean Skies Foundation, April 2013.

[74] U.S. Patent Appl. 20140018581 (July 8, 2013) to Siluria Inc.

[75] U.S. Patent Appl. 20140012053 (January 9, 2014) to Siluria, Inc.

[76] U.S. Patent Appl. 20120116137 (May 10, 2012) to Primus Green Energy.

[77] Chemical & Engineering News, April 30, 2012, p. 34.

[78] Automotive World, January 20, 2014.

[79] Guiness World Records, January 22, 2013.

[80] Toyota Europe News, July 3, 2013.

[81] Alternative Fuels Data Center: U.S. Department of Energy.

[82] Chemical & Engineering News, April 30, 2013, p. 32.

[83] Chemical & Engineering News, December 13, 2013, p. 11.

[84] Wikipedia, The Free Encyclopedia. http://en.wikipedia.org/Wiki/File:Solid_oxide_fuel_cell. svg, May 2011.

[85] J. Spendlow, J. Marcinkowski, Fuel Cell System Cost, Dept. of Energy Fuel Cell Technologies Office, 2013.

[86] Fuel Economy: Where the Energy Goes: U.S. Dept. of Energy. <www.FuelEconomy.gov/ feg/atv.shtml>, 2011.

[87] USA. Today, October 2, 2013.

[88] Chemical & Engineering News: October 7, 2013, p. 24.

[89] The New York Times, March 16, 2014.

[90] <www.energy.gov/eGallon> U.S. Dept. of Energy.

[91] The New York Times, May 13, 2014, p. B3.

[92] C. Walcott, E.B. Rifkin, Ind. Eng. Chem. 43 (1951) 2844.

[93] E.B. Rifkin, Proc. Am. Petrol. Inst. III 38 (1958) 60.

[94] Long Island Utility Fighting To Defeat MTBE Safe Harbor; Napoli, Kaiser, Bern & Associates, March 16, 2004.

[95] ICIS News, July 5, 2006.

[96] L. Varo, International Energy Agency, December 29, 2013, Medium Term Coal Rept.

[97] China Greentech 2012; May 14, 2012. <http://ukmediacentre.pwc.com/news-release>.

[98] USA Today, April 20, 2014.

[99] U.S. Energy Information Administration, Annual Energy Outlook, December 16, 2013.

[100] Green Car Cong., April 18, 2014; ExxonMobil Outlook For Energy; A View to 2040.

[101] Biodiesel Magazine, April 29, 2014.

[102] Today In Energy; U.S. Energy Information Admin., May 2, 2014.

[103] Annual Energy Outlook 2014, U.S. Energy Infor. Admin, May 7, 2014.

[104] Arizona Republic: March 31, 2014; p. A 29103.

[105] A. Demitribas, M.F. Demitribas, Energy Convers. Manage. 52 (2011) 163.

INDEX

Printed in the United States
By Bookmasters